# NETWORKS IN SOCIAL POLICY PROBLEMS

Network science is the key to managing social communities, designing the structure of efficient organizations and planning for sustainable development. This book applies network science to contemporary social policy problems.

In the first part, tools of diffusion and team design are deployed to challenges in adoption of ideas and the management of creativity. Ideas, unlike information, are generated and adopted in networks of personal ties. Chapters in the second part tackle problems of power and malfeasance in political and business organizations, where mechanisms in accessing and controlling informal networks often outweigh formal processes. The third part uses ideas from biology and physics to understand global economic and financial crises, ecological depletion, and challenges to energy security.

Ideal for researchers and policy-makers involved in social network analysis, business strategy, and economic policy, the book deals with issues ranging from what makes public advisories effective to how networks influence excessive executive compensation.

BALÁZS VEDRES is Associate Professor and Director for the Center for Network Science, Central European University. His research is focused on understanding historical dynamics in network systems.

MARCO SCOTTI is Principal Investigator at The Microsoft Research – University of Trento Centre for Computational and Systems Biology (COSBI), where he researches in molecular nutrition, biological networks, and stochastic simulations.

# NETWORKS IN SOCIAL POLICY PROBLEMS

*Edited by*
BALÁZS VEDRES
*Centre for Network Science, Central European University*

MARCO SCOTTI
*The Microsoft Research University of Trento Centre
for Computational and Systems Biology*

CAMBRIDGE UNIVERSITY PRESS
Cambridge, New York, Melbourne, Madrid, Cape Town,
Singapore, São Paulo, Delhi, Mexico City

Cambridge University Press
The Edinburgh Building, Cambridge CB2 8RU, UK

Published in the United States of America by Cambridge University Press, New York

www.cambridge.org
Information on this title: www.cambridge.org/9781107009837

First published 2012

*A catalogue record for this publication is available from the British Library*

*Library of Congress Cataloging in Publication data*
Networks in social policy problems / edited by Balázs Vedres and Marco Scotti.
p. cm.
Includes bibliographical references and index.
ISBN 978-1-107-00983-7 (hardback)
1. Social problems. 2. Social networks. 3. Applied sociology.
4. Social sciences–Network analysis. I. Vedres, Balázs. II. Scotti, Marco.
HN18.3.N48 2012
301–dc23    2012016088

ISBN 978-1-107-00983-7 Hardback

# Contents

# Contributors

**Stefano Allesina**
*Department of Ecology and Evolution, Computation Institute, The University of Chicago, Chicago, USA.*

**Albert-László Barabási**
*Center for Complex Network Research and Departments of Physics, Computer Science and Biology, Northeastern University, Boston, USA.*

**Sandrine Blanchemanche**
*INRA-Met@risk, Paris, France.*

**Antonio Bodini**
*Department of Environmental Sciences, University of Parma, Parma, Italy.*

**Cristina Bondavalli**
*Department of Environmental Sciences, University of Parma, Parma, Italy.*

**Stephan Clemençon**
*Institut Telecom – Telecom Paristech, Paris, France.*

**Noshir Contractor**
*Northwestern University, Evanston, USA.*

**Peter Csermely**
*Semmelweis University, Department of Medical Chemistry, Budapest, Hungary.*

**Charanpal Dhanjal**
*Institut Telecom – Telecom Paristech, Paris, France.*

**Mario Diani**
*ICREA – Universitat Pompeu Fabra, Barcelona, Spain.*

**Meikuan Huang**
*Northwestern University, Evanston, USA.*

**Yun Huang**
*Northwestern University, Evanston, USA.*

**Benjamin F. Jones**
*Northwestern University, Kellogg School of Management, Evanston, USA.*

**Ambrus S. Kaposi**
*Semmelweis University, Department of Medical Chemistry, Budapest, Hungary.*

**Bruce Kogut**
*Sanford C. Bernstein & Co. Professor, Columbia Business School, Columbia University, New York, USA.*

**István A. Kovács**
*Department of Physics, Loránd Eötvös University, Budapest, Hungary.*

**Emmanuel Lazega**
*Department of Sociology, University of Paris-Dauphine, Paris, France.*

**Drew Margolin**
*University of Southern California, Los Angeles, USA.*

**Ágoston Mihalik**
*Semmelweis University, Department of Medical Chemistry, Budapest, Hungary.*

**Lise Mounier**
*CMH-CNRS, Paris, France.*

**Tibor Nánási**
*Semmelweis University, Department of Medical Chemistry, Budapest, Hungary.*

**Katherine Ognyanova**
*University of Southern California, Los Angeles, USA.*

**Robin Palotai**
*Semmelweis University, Department of Medical Chemistry, Budapest, Hungary.*

**Ádám Rák**
*Robotics Laboratory of the Faculty of Information Technology, Péter Pázmány Catholic University, Budapest, Hungary.*

**Akos Rona-Tas**
*University of California, San Diego, La Jolla, USA.*

**Fabrice Rossi**

*Institut Telecom – Telecom Paristech, Paris, France.*

**Marco Scotti**

*Center for Network Science, Central European University, Budapest, Hungary. The Microsoft Research – University of Trento Centre for Computational and Systems Biology (COSBI), Rovereto, Italy.*

**Jarrett Spiro**

*INSEAD Organisational Behaviour, Singapore.*

**David Stark**

*Department of Sociology, Columbia University, New York, USA.*

**Máté S. Szalay-Beko**

*Semmelweis University, Department of Medical Chemistry, Budapest, Hungary.*

**Zoltán Szántó**

*Institute of Sociology and Social Policy, Corvinus University of Budapest, Budapest, Hungary.*

**István János Tóth**

*Institute of Economics, Hungarian Academy of Sciences, Budapest, Hungary.*

**Brian Uzzi**

*Northwestern University, Northwestern Institute on Complex Systems (NICO), Evanston, USA.*

**Szabolcs Varga**

*Institute of Sociology and Social Policy, Corvinus University of Budapest, Budapest, Hungary.*

**Balázs Vedres**

*Center for Network Science, Central European University, Budapest, Hungary.*

**Douglas R. White**

*Department of Anthropology, University of California, Irvine, USA.*

**Stefan Wuchty**

*Northwestern Institute on Complex Systems (NICO), Evanston, USA.*

**Jae-Suk Yang**

*Associate Research Scholar, The Sanford C. Bernstein & Co. Center for Leadership and Ethics Columbia Business School, Columbia University, New York, USA.*

# Acknowledgements

The editors would like to thank Central European University for a seed grant that helped to bring network researchers together to reflect on social policy problems. We are grateful to Ferenc Jordán for his help in organizing the conference entitled "The unexpected link – Using network science to tackle social problems" (Budapest, June 17–18, 2009). This conference served as founding event for the Center for Network Science (Central European University) and inspired the birth of this book. Bianca Baldacci is kindly acknowledged for her graphic design contribution. She conceived and realized the cover photograph, and provided fundamental support in the optimization of the figures. The editors are grateful to Melissa J. Morine and Silvia Neamtu for their useful comments. Andrea Scotti, Tommaso Schiavinotto, and Sehar Tahir are acknowledged for their contribution in formatting the Latex version of the book. Special thanks are dedicated to Simon Capelin, Lindsay Barnes, and Antoaneta Ouzounova, the team at Cambridge University Press. Last but not least the editors are grateful to all authors who contributed to the book: their help and patience were fundamental to completion of this volume.

# 1

# Introduction

BALÁZS VEDRES AND MARCO SCOTTI

## 1.1 Introduction: applied network science

This volume provides an overview of network science applied to social policy problems. Network science is arguably the most dynamic and interdisciplinary field that has grown up to address problems of an increasingly interconnected world. Social problems transgress disciplinary boundaries, especially with the ever-increasing complexity of our globally interconnected world society. It is natural that with complex problems such as loss in ecological diversity, economic crisis, spread of epidemics, and the safety of our food supply, we turn to a field of research that focuses on explaining complex dynamics, and that is inherently interdisciplinary itself.

Networks have become part of our everyday experience as we routinely use online social network services, we hear reports on the operations of terrorist networks, and we speculate on the six degrees of separation to celebrities and presidents. Less manifestly, we rely on vast and complex infrastructural networks of electric power distribution, Internet data routing, or financial transfers. We only ponder the complexity of these systems when we are faced with avalanche-like dynamics in their collapse, as major blackouts, system stoppages, or financial meltdowns.

With networks on the collective mind, there is ample interest in tools to understand and manage complex network systems of social ties. At the time of writing this introduction (in October 2011), there were eighteen applications available on Facebook to visualize one's social network. The popularity of such software tools shows our fascination with the interesting new perspective that the graphic visualization of friendship ties provides. While many of these little apps are for the purposes of entertainment, high-power software platforms are being put in place to assist scientific communities, business work teams, or emergency response teams to efficiently map out not only their social network neighborhood, but also who possesses critical

Networks in Social Policy Problems, eds. Balázs Vedres and Marco Scotti. Published by Cambridge University Press © Cambridge University Press 2012.

pieces of information, tacit knowledge, or expertise in their vicinity (Contractor, 2009). Network analysis has become an essential tool in the counter-insurgency battlefield. As the enemy is increasingly organized into non-hierarchical networks, army operatives need to learn methods to visualize and measure the weak points of the insurgency network (Bunker, 2006). In the capture of Saddam Hussein and Osama Bin Laden, a crucial element was a social network analysis approach that helped to sort relevant ties and key couriers from the vast pool of intelligence data.[1]

Network science tools are also being deployed to understand and manage biological or engineered systems. With the increasing reliance on green energy, and the depletion of fossil fuels, electric energy is becoming ever more prominent, increasing the load on the power grid. Network science tools help in the design of power grids that are more resilient to failure and attack (Holmgren, 2006; Pagani and Aiello, 2011). With the steady decline in fisheries productivity, there is increasing attention on the fragility of marine food webs. The simulation tools in Ecopath and Ecosim have been developed to understand the impact of policy and economic changes (such as the size of fishing vessels or the size of fishing conglomerates) on marine ecosystems from a network perspective (Pauly *et al.*, 2000). Such tools can help avoid fishing to the point where catastrophic collapses happen in food web networks (Espinosa-Romero *et al.*, 2009). Network models are being adopted in cancer research as well. The prediction of cancer metastasis connections (Chen *et al.*, 2009), and the better understanding of the metabolic links of cancer cells (Jeong *et al.*, 2000) all contribute to a more targeted treatment of cancer.

Behind these tools there is the new science of networks, an interdisciplinary field that aims at explaining complex phenomena at larger scales, emerging from the simple principles of making network links locally. Witnessing a decade of explosive growth, academics from the natural and social sciences are engaged in basic research to unveil principles that govern network behavior in a wide range of empirical fields. The power of this field lies in its strong and diverse roots: in the social sciences, attention to the social web dates back to the works of Georg Simmel, from the last decade of the nineteenth century (Simmel, [1922] 1964). Drawing graphs of social ties (sociograms) became a key tool of psychology in the 1930s through the work of Jacob L. Moreno, who famously aspired to draw the social network graph of the entire New York City, and got as far as drawing a graph of 435 individuals (Moreno, 1934).[2] Attention to networks was revived using computing power in the sixties and seventies, to understand social structure as blockmodels

---

[1] On May 12, 2011 the BBC news featured Dan Pearson from a company named i2, who presented the capabilities of their social network reconnaissance software system in a video interview (http://www.bbc.co.uk/news/technology-13336281).

[2] A high quality digital version of the original sociogram of 435 nodes is available here: http://www.asgpp.org/docs/WSS/WSS.html

(White *et al.*, 1976). By the nineties social network analysts had developed rich methods to estimate individual network centrality and power, role structures, balance, and structural holes. In the natural sciences the roots of network thinking reach back to the beginnings of graph theory, to the work of Leonhard Euler in the eighteenth century. Research on ecological networks dates back to the pioneering work of the Italian zoologist Lorenzo Camerano (1880), who constructed the earliest known graph of feeding relations. The theory of random graphs of Paul Erdős and Alfred Rényi from 1959 was crucial in catalyzing thinking about complex systems. Within the physics of condensed matter, studies of spin glasses in the seventies drew attention to the complexity arising from the stochastic interactions of particles. In biology, increasing attention was paid to the complex network dynamics of metabolic networks and protein interactions.

While the roots of network science are diverse, there has been frequent borrowing amongst disciplinary, methodological, and thematic groups throughout its history. Network science entered a new phase of integration during the first decade after the turn of the millennium. Scholars and research centers emerged following research agendas that truly integrated disciplines. The field of network science in its recent phase of evolution maintains a network-like structure that continually bridges and bonds disciplinary groups, avoiding compartmentalization, yet consolidating a common language and methodological toolbox. For example, protein–protein interaction models in biology borrowed from sociological metrics of betweenness centrality (Hahn and Kern, 2005). Or, social movement studies adopted methods directly from the physics of phase transitions (Oliver and Myers, 2003). The key researcher who best exemplifies the new science of networks is Albert-László Barabási. His work on scale-free network structures, the dynamics of events unfolding in networks, and the possibilities of controlling networks attracted tens of thousands of citations. His readership spans disciplinary boundaries, and his center brings together researchers with diverse backgrounds.

A shared feature of all network research is the use of innovative, experimenting, highly visual methods. Methods in network science are highly modular. As a contrast, consider linear regression (or any regression-like method): such methods force the structure of the technique on the whole of the research project. Our metaphor is a ten-ton highly specialized machine tool. Once you buy it, you have to shape your workshop to suit the inputs of the machine, hire workers with specialized skills, and find specific buyers for the output. While there are fixes and modifications to regression methods, the basic logic stays the same. In contrast, network methods are built modularly from general purpose tools, and allow great flexibility. One can take an algorithm developed in anthropology to identify typical kinship arrangements (like, for example, "blood marriage" – the marriage of cousins) and take this pattern search idea to citation network research, where instead of parents and descendants, there are citing and cited works. The notion of a certain lineage pattern can be used to identify emerging disciplinary schools, rather than kinship clans. Then one can augment the kinship pattern search algorithm by a statistical evaluation of pattern frequencies.

With recent advances in the availability of computing power, ideas about complex network dynamics are being tested with large datasets. The enhanced potential of high-speed computation has opened up new vistas to take on previously unimaginable challenges. With a large number of nodes, qualitatively different dynamics can be observed, similarly to the notion of thermodynamic limit in physics (with the increase in the number of particles, one can switch from conceptualizing colliding particles to the notion of gas). Visualization has always been a key element in network science. An efficient map of network topography is a great tool to show group divisions, central players, and main directions of flow. At websites like www.visualcomplexity.com, vast collections of visualizations are available, that border on an art form. Again, the idea is that complexity can be grasped given efficient tools.

While the conceptual roots and the applications are diverse, the fundamental principles of network science are shared. What is distinctive about the conceptual foundations of network science? What is, and what is not a case of networked phenomena? While taking a network approach is productive for many problem areas, one should refrain from seeing networks everywhere. It seems that the concept of network is being evoked all too casually these days: there is much talk of social networking, the old boy network, or a network society. It seems that networks are everywhere, but network science uses this concept with a careful definition. What is a network in a network science sense? The mere possibility of identifying nodes and drawing lines between them is different from identifying a network mechanism that explains complex outcomes from simple building blocks. At a railway station, for example, one could identify travelers as nodes, and draw ties between any two of them if they are headed to the same destination, or, if they purchased their ticket from the same teller. These ties, however, would hardly figure as part of a theoretically sound explanation for the behavior of individuals, or for the functioning of the transit system. However, if you drew a network where destinations are nodes, and the number of travelers moving between them is the tie, you can start to identify mechanisms of systemic breakdown in the case of a major accident.

While nodes and ties are the fundamental building blocks to any network science approach, the power of network science lies in focusing on triads and beyond: the broader topographical context. A connectionist perspective (focusing on dyads, pairs of nodes) does not exhaust the possibilities that a network perspective can offer. Recognizing that connections matter is only the first step: recognizing that ties around dyads matter begins to unleash the power of network science. A tie to a political party might matter for a firm to have access to resources and timely information on policies. Thus far this is a question of connections – is a firm connected to a party? But one can also ask: once a firm is connected to a political party, how are its ties to other firms affected? This becomes a network question: a tie formed at one part of the network system influences the probability of ties elsewhere, and might help explain the emergence of a politically polarized economy (Stark and Vedres, in press).

While this all might seem rather abstract, network science offers a fresh, simple set of heuristics to tackle practical problems around us. When we speak of complex dynamics, we think of processes that are beyond the reach of common sense. For example, the cause of a massive wave of demonstrations with several thousand participants (such as those in the Middle East over the spring of 2011) might elude our common sense if the timing of the demonstration shows no correlation with the timing of grievances. Why now, why so many? A network science approach would start from the ties between activists, and the way they convince each other to participate. Assuming the simplest process of peer pressure ("I go if at least two of my friends go"), the resulting process shows complex dynamics akin to an avalanche: massive participation is sudden, following a long period of gestation (Kim and Bearman, 1997; Oliver and Myers, 2003).

While network science can appear complex, even esoteric at first, it is readily applicable to a wide range of social problems. Instead of providing complex answers to simple questions, network thinking offers new heuristics, based on fundamental and simple, yet startling regularities. Our common sense in the subjective assessments of probabilities predominantly rests on the assumption of unconnected units (cells, human individuals, businesses, states), and their autonomous decision-making (Tversky and Kahneman, 1974). Such heuristics lead us astray in many problem-solving situations where dynamics are influenced by connected agents. As we show in this volume, the range of such problems is truly wide, from team creativity to compensation of executives, to financial crises and ecological disasters. As a start let us consider three examples for heuristics based on the assumption of connections and interplay that one should adopt from network science.

The first example of such heuristics is the idea of a *small world*. We generally assume that a new epidemic outbreak in the heart of Africa or a new mobile phone virus spreading in Taiwan are too far and insulated from us. In reality, our social networks are much more connected than we would prefer. While many of our ties are within closely knit clusters of friends and colleagues, there are enough long distance ties in our network vicinity, so that there is a short path starting somewhere near you to any part of the world. The concept of a small world was coined from the experience of, as we discover, the shared acquaintances with a random person we happen to sit next to on a long flight – what a small world! Duncan Watts, the researcher who first described this phenomenon in mathematical terms, showed that social networks occupy a very narrow range between order and randomness, where local clustering is high (there are dense pockets of friends), while overall path lengths are low (there is close reachability globally). It is much more probable to catch a new strain of flu, or to have your smart-phone turned into paperweight by a virus infection than you would assume. This research insight is now an integral part of epidemic forecasting (Pastor-Satorras and Vespignani, 2001).

A second example for a network heuristic is that there are no typical participants: networks are *scale-free*. Social and technological networks are scale-free; there is not a particular peak in network embeddedness, such that at any scale, inequalities show a similar distribution. This means that there are huge inequalities in network systems, such as the World Wide Web. While naïve approaches would assume that "getting online" is the decisive step to transgress the digital divide, the distribution of incoming links – the main channels of attention – is extremely unequal (Barabási and Albert, 1999). Getting online is just the first step, and for most website startups the investment into hardware, programming skills, and graphic design brings no visibility whatsoever. Most websites have no links pointing to them at all, while there are very few that attain millions of incoming links. In the world of webmasters making links, but also in the realm of choosing new friends or making new business partnerships, the winner seems to take all: links are made to participants with many links already. This is true for our social networks as well. Instead of most people having an average number of friends, with a few isolated and a few well connected individuals, most people have a relatively low number of friends, while there are several well connected individuals, then there are enough extremely well connected individuals, and even more than a few that are unbelievably well connected. This means that success in a connected world takes persistence: the first links are difficult to attract, but then there are increasing returns to scale: having more friends makes it much easier to gather more friends still. This also means that the robustness of social and technical networks hinges on a few key hubs, which highly limits their tolerance to directed attack.

The third example of a heuristic from network science is about disproportionate sizes of causes and effects. Processes of peer pressure and imitation, the spread of innovations, or the collapse of energy delivery systems all unfold in a non-linear way. If something becomes fashionable, suddenly almost everyone seems to carry that item (a new gadget or a new piece of clothing). This contradicts our common sense assumption that the outcome of a process is linearly proportional to the amount of energy invested – that large events need big causes. We assume that a particularly popular brand of shoe must be a result of an expensive marketing campaign. Charles Tilly was among the first to point out how deeply this heuristic is rooted in our conceptions of social change (Tilly, 1984): historians assume that large revolutions are ignited by the gravest of grievances, for example. What network science found, instead, is that when social movements are near critical phase transition points, a small increase in activism can lead to massive protests and revolutions. In the case of engineered systems we are more accustomed to the notion that a small fault can lead to a major collapse, as in the case of the 2003 blackout of North America (Albert *et al.*, 2004). In the case of ecological systems, it is much less part of public consciousness that beside protecting charismatic species (that are easy to stand up for, such as whales or turtles) one also needs to pay attention to the keystone species (possibly a small crustacean). Keystone species show disproportionate effects on ecosystems in comparison with their relative abundance. Such species are at key positions in the ecological network, and losing them can set off major avalanches of ecological collapse (Ulanowicz, 2009).

## 1.2 The structure of this volume

This volume provides practical examples along these lines, arranged in three parts that correspond to the three network heuristics outlined above. The first part of the book focuses on the construction and structure of social networks among individuals, and how the structure of such networks influences information flows, and the emergence of new ideas. The heuristic of a small world connects the typical structure of social networks with cohesive groups and occasional long distance ties with the speed and dynamics of spreading information and epidemics. This part further elaborates on the impact of our social networks on the flow of ideas. Chapters in this part offer insights into how the decision to pass on ideas is conditioned by properties of network ties, and how network structures help in generating ideas in the first place.

Chapter 2 focuses on what predicts spreading information in a social network, using the small-world network approach. While counterintuitive at first, compared with mass media, social network ties can be more effective channels of information. Mass media can reach more people faster with a message but it is less likely that the message will be acted upon. This is of especially grave concern for public advisories and marketing campaigns. A message that spreads via a social network tie – passed on by an acquaintance – can be customized for the receiver, can be told in the right context and in the right moment and, most importantly, can be endowed by the trust that the receiver has in the sender. These information transmission ties, however, need to be placed in the network context of other ties. Depending on the network location where information starts spreading, the diffusion process might reach almost everyone fast, or it might peter out on the peripheries.

Networks are effective channels for information flow, but can they also contribute to the generation of new ideas? Chapter 3 focuses on teams in creative fields – science and Broadway musical production – to identify the significance of network connections in success. The prevalence of teamwork in creative production is clearly increasing: new ideas are increasingly products of connected individuals rather than solitary geniuses. Teams themselves can be more or less successful and this is a function of their position in a larger collaborative network. The most successful teams are composed of a mix of incumbents with trusted and tried collaborative ties, and newcomers bringing fresh ideas and access to as yet untapped areas of the network. Thus, again, the dual nature of our social networks leaves clear marks: we need cohesive groups to produce ideas, but these groups are most efficient if they make use of the long reaching ties, to be embedded in the broader flow of ideas.

Chapter 4 identifies a network structure that is closely associated with the generation of new ideas. The focus here is on the network structures that enable novel insights. Truly innovative ideas – in the first instance, a fresh conceptualization of the problem itself – are not free-floating outside social groups. Innovation is a generative recombination that requires familiar access to diverse resources. The chapter identifies a network structure where dense regions of ties overlap – a structural fold – that facilitates such recombination. There seems to be a tradeoff for social groups: maintain exclusivity for long-term stability with modest performance, or be open to overlaps for high performance and the risk of disintegration. The results of this chapter indicate though that business groups have devised ways to maintain stability while also reaping the innovation benefits of structural folding. Coherence of larger business groups – a business group proper – was achieved by a historical lineage of group breakup, merger, and reunification. Successful business groups seem to have an inner network structure with overlapping subgroups that are frequently re-shuffled to maintain a steady production of fresh ideas.

Chapter 5 focuses on the processes of composing a team in the context of nanotechnology, and potential entry points for policy-makers to intervene in team composition. While the preceding chapters aimed at uncovering the predictors of efficiently spreading and successfully inventing ideas, this chapter makes a further step to ponder the path by which the team is assembled. Successful teams possess both a favorable mix of attributes (along gender for example) and include stars with many ties at a broader network scale. A team brings together a mix of career histories and personal and social categories. From a policy-maker's perspective the question is: what are the factors that one can influence, encourage, or discriminate to help create better teams? A key balancing act here is not to stifle the power of self organization, while maintaining the tools that guide team formation. The key to successful intervention is knowledge about the emerging team network structure: the policy-maker can identify patterns and principles that none of the participants on the ground sees.

The second part broadens the conceptual horizon to organizations and institutions, and how network structures influence power and the abuse of power in an organizational context. In line with the heuristic of scale-free networks, the attention in this part is on inequalities of power and money. This part offers a shift of perspective: instead of the circuitry metaphor inherent in the small-world approach to the creation and spreading of ideas, this part focuses on power fields. Networks are far from being the domain of homogeneity and equity. To understand organizational structures of social movements, inequalities in CEO compensation, mechanisms of organizational capture, or the workings of political corruption, this part offers a look at power in networks.

Chapter 6 introduces the idea of organizational fields, focusing on the workings of social movements. The first question that arises when one tries to understand collaboration and power in a network, is where the boundaries of the network lie, and how these boundaries get defined. While organizations have natural boundaries, the perimeter of a collaborating network of organizations is much more blurred. The internal structure of social movement networks is strongly related to mechanisms of coordination. For example, denser, more horizontal network structures spend more energy on deliberation, while centralized network structures are more efficient in coordination; but they are also more fragile and dependent on central nodes.

Chapter 7 takes the tools of organizational network analysis to explain the size and structure of executive compensation. Compensation is supposed to provide incentives for executives, such that the executive (the agent) follows the priorities of shareholders (principals), in a complex landscape of decisions of which the principals have limited oversight. Principals are typically concerned with too little risk taking by agents, thus compensation packages include a large portion of a non-salary (equity) element. For the policy-maker the problem is the reverse: how to intervene and shape executive compensation such that excessive risk taking would be curbed. If financial firms are ranked according to CEO compensation in 2006, we observe that only one of the top five survived the financial crisis of 2008. To understand entry points for the policy-maker into shaping executive compensation, one needs to know the mechanisms by which companies benchmark their compensation packages. The strongest elements are the interaction of the peer network (the other companies that a given company chooses to follow) and the mobility network of CEOs. A CEO seduced from outside of the company usually enjoys a substantively higher compensation package, which then sets off cascades in the benchmarking network.

Chapter 8 introduces the notion of power and capture into such organizational fields. Powerful private actors endowed by resources and network centrality can capture public organizations, as in the case of the Commercial Court of Paris. Bankers managed to occupy key positions in this court, that originally aimed to exercise control over them; while finance represents only about five percent of the constituency of the court, almost a third of the judges came from the financial sector.

The explanation of this process of capture is in the network position of bankers: they occupied an increasingly central role in the informal advice network of judges. As the selection of new judges is by invitation from currently serving judges, the representation and centrality of bankers in this organization was increasing steadily.

As Chapter 9 shows, organizational networks can be misused for private gain. Corruption is essentially a parasitic use of organizational network resources; it is inherently a network phenomenon, where trust and covert resource movements are channeled through informal network ties. If we consider network patterns of corruption, organizational networks are integrated with covert personal networks, hidden principals, brokers and quasi-clients. Network embeddedness can be analyzed referring to four types of corruption schemes: bribery, extortion, embezzlement, and fraud. As examples from Hungary show, the smooth operation of these schemes relies on organizational power inequities, lack of transparency, exposure, and vulnerability.

The third part of the volume turns to dynamics at a larger scale, with cases of complex dynamics at the level of whole systems – biological, economic, ecological, and engineered. Chapter 10 brings lessons learned from biological crisis at the cellular level to systemic crises in the current financial meltdown. Interactions in yeast and human cells under stress become more modular, with a decrease in network ties across segments. The exceptions are stress proteins that maintain their diverse and far reaching portfolio of network ties. In the recovery of the cell after crisis, these proteins take the lead, and define the shape of the new network. The lesson for financial systems and social networks in general is that we can expect a decrease in connectedness (fewer transactions, an increased internal focus on organizations), and that those who can maintain a broader network during the crisis, will become key players as the system bounces back and reconstructs itself.

Chapter 11 takes a broad historical perspective on globally interconnected economic networks. Starting from early historical examples of global trade network systems (the silk road from the tenth to the twelfth century, Champagne fairs from the thirteenth to the fourteenth century, the silk road from the thirteenth to the fourteenth century, and East India trading in the seventeenth century), a mode of operation for global economic networks based on a code of transparency is outlined. The dynamics of the global system becomes markedly different when a new code of operation emerges based on guardian ethics. The global system with European dominance that emerged from the mid fifteenth century – based on financial capital backed by military power – produced hyper-inequalities, and a dynamics with deeper crises. Policy lessons are gathered from the alternative logics of earlier global systems.

Chapter 12 introduces network science tools to understand the functioning of ecological systems. As the stress from human activities on natural habitats is increasing exponentially, it is vital to capture the non-linear way in which these systems might undergo stress events, eventually leading to their collapse and the extinction of the species. By analyzing the fragility of ecological networks and studying their dynamic behavior, strategies for the management of natural systems can be implemented. Interventions can target keystone species: species that might not be at the focus of attention otherwise, but when lost large parts of the ecosystem would collapse. Network science also provides a better understanding of the thresholds beyond which ecosystem damage might become irreversible.

Chapter 13 takes tools developed in ecosystem network analysis to understand resilience and collapse in energy delivery infrastructures. By analyzing the European natural gas pipeline system, an alarming portrait emerges. Concerns of supply security become increasingly dependent on network structure, rather than being a simple function of the distance from natural gas sources. While multi-billion Euro investments into infrastructures will mitigate somewhat the European dependence on Russian sources, they will increase network-based risks. The energy delivery system is evolving to be more fragile, with more and more politically unstable states gaining influence on the energy security of European countries.

Chapter 14 concludes the volume by providing an overview on the state of network science today, and directions for the future. Starting with the example of graphite, diamond, and nanotubes – all are substances made up of the same carbon atom and the same type of chemical bond – the chapter reflects on the many ways in which awareness of network structures shapes strategic vision. With recent advances in methods and data processing, and recent challenges where complexity became ever more apparent, network science holds the promise to inform policy-making in a wide range of social problems.

# Part I

Information, collaboration, innovation: the creative power of networks

# 2

# Dissemination of health information within social networks

CHARANPAL DHANJAL, SANDRINE BLANCHEMANCHE,
STEPHAN CLEMENÇON, AKOS RONA-TAS, AND FABRICE ROSSI

## 2.1 Introduction

State agencies responsible for managing various risks in social life issue advisories to the public to prevent and mitigate various hazards. In this chapter we will investigate how information about a common foodborne health hazard, known as Campylobacter, spread once it was delivered to a random sample of individuals in France. The Campylobacter is most commonly found in chicken meat and causes diarrhea, abdominal pain, and fever. The illness normally lasts a week but in rare cases patients can develop an auto-immune disorder, called Guillain–Barré syndrome, that leads to paralysis and can be deadly. Campylobacter, together with Salmonella, is responsible for more that eighty percent of foodborne illnesses in France and strikes over 20,000 people each year. People can take simple steps to avoid infection by cleaning their hands, knives, cutting boards, and other food items touched by raw chicken meat and by cooking the meat thoroughly.

In this chapter we build two different network models to see how the information about Campylobacter diffuses in society, by mapping onto various network structures the data we gathered with three waves of surveys. In these models the spread of information depends on two sets of factors. Firstly, each person has a set of individual properties that influences their propensity to transmit the information to or to receive the information from someone they know. Second, each person is connected to others in ways that also affect transmission. There are three aspects of these social ties that matter. As the information travels through existing ties, the quantity of connections should have an influence, as people with more ties should have more opportunity to disseminate the information. The quality of ties should matter as well, because certain types of ties may be more conducive to information

Networks in Social Policy Problems, eds. Balázs Vedres and Marco Scotti. Published by Cambridge University Press © Cambridge University Press 2012.

transmission than others. Finally, the overall structure of the entire network, i.e. how egocentric networks are linked into a larger whole, should also play a role. Our surveys provide data for the individual characteristics, as well as the quality and quantity of the social ties. In the diffusion model, two different overall network structures, the Erdős and Rényi random, and the small-world (SW) model, are introduced through modeling assumptions.

The central question of this chapter is how individual characteristics and the various aspects of social networks influence the spread of information. A key claim of our chapter is that information diffusion processes occur in a patterned network of social ties of heterogeneous actors. Our percolation models show that the characteristics of the recipients of the information matter, as much if not more than the characteristics of the sender of the information, in deciding whether the information will be transmitted through a particular tie. We also found that, at least for this particular advisory, it is not the perceived need of the recipients of the information that matters but their general interest in the topic.

As for the diffusion of information, we found that the two network structures behave differently in some ways. If the proportion of the population that receives the information initially from the center (our survey) is lower, the random graph model diffuses the information to a larger segment of the population then the SW model, since the advisory travels farther in random networks. However, as the initial exposure increases above a certain level, the two models deliver the advisory by word of mouth to very similar proportions of the population. For the SW models we find that as the size of the group that is initially exposed grows, diffusion first increases then decreases and there is an optimal proportion of the population that the message initially has to reach to result in the maximum word of mouth diffusion. Since the initial deployment of the information costs money and the subsequent diffusion is costless for the center, finding this optimal size is of practical importance.

Finally, we offer a visual presentation of the transmission process and the way the composition of those receiving the message shifts. Our analysis finds that distribution of particular characteristics of message recipients changes most in the first round, and in later rounds it gravitates towards the distribution of the characteristics in the population.

## 2.2 Theoretical overview

### 2.2.1 Diffusion of health information and interpersonal communication

Major public health campaigns are aimed at modifying individual conduct, by either trying to decrease risky behaviors, such as alcohol abuse and smoking, or seeking to increase health promoting activities such as exercise and following a healthy diet (Compas *et al.*, 1998). The approaches to the dissemination of advisories in the field of public health have varied across time, countries, and objectives of the campaigns, but they are systematically based on specific views of society and on certain assumptions about social processes.

*Broadcast approach*

The dominant approach in disseminating health advisories has been the broadcast method, where the information is released centrally and it is targeted at members of the public individually. The main assumption behind the broadcast approach is that the message travels directly from the center, usually the health agency, to each citizen via various channels of mass communication. The success of the broadcast approach depends on reaching as many people as possible with the broadcast, and on delivering the message in ways such that the desired effects are created in each member of the audience. The broadcast model is a hub-and-spokes system with the agency in the middle and each unconnected member of the public at the end of one spoke. Communicators using this model, therefore, concentrate on two aspects of the process: reach and stickiness. Reach is the matter of getting the message to the largest possible part of the target audience and involves careful planning of the dissemination of the information. Communicators must decide which media are best suited to their purposes, where they should deploy the information, when, and how many times. Stickiness is a matter of creating the proper effect once the information is delivered. Here, the main objective is getting people to pay attention to, understand, and act on the information in question. To improve individual reception of their messages, broadcasters have spent a lot of effort trying to understand the psychology of message reception and perception in their quest to craft the most effective messages.

### 2.2.2 The two-step model
*Political mass communication*

The indisputable strength of the broadcast model is that it provides the maximum control for the center over the message, as each recipient obtains the information from the central source. Doubts about the broadcast model, nevertheless, were raised early on. One of its first critics was the sociologist Paul F. Lazarsfeld. In a series of studies on political communication in the 1940s he found that the direct effect of broadcasted political messages is negligible and that most people acquire their political opinions not from television, radio, or newspapers, but from other people he named "opinion leaders." He argued that political messages produce their effects – if at all – in a two-step process: the original message is picked up by the opinion leaders who then pass it on to the rest of the community (Lazarsfeld *et al.*, 1948; Katz and Lazarsfeld, 1955). Lazarsfeld's main insight was that most people receive political messages not from the media, or from a central point of emission, but by word of mouth from others in their community and therefore, understanding the structure of the community is crucial for making communication effective (Katz, 1996; Burt, 1999).

## Marketing

A similar idea surfaced in market research where advertisers encountered the same difficulty in getting their messages across. Researchers found that just as citizens in political discourse, consumers often obtain their information through social ties (Katona and Mueller, 1955) and not from the advertisements with which they are bombarded. Following this observation, the field of market research has distinguished three types of customers who influence others: the early adopter, the opinion leader, and the market maven. Early adopters of new products influence others by buying the product. Their purchases inform other people that the product is available and worth acquiring. Opinion leaders have special expertise about a particular piece of merchandise and dispense it to those interested in the product. Market mavens, on the other hand, are individuals who research and plan their purchases and pay a lot of attention to getting the best deal, and as a result, "have information about many kinds of products, places to shop, and other facets of markets, and initiate discussions with consumers and respond to requests from consumers for market information" (Feick and Price, 1987; Clark and Goldsmith, 2005). Marketing people are eager to find early adopters, opinion leaders, and market mavens. They focus on the special characteristics of these senders of information. If they can sell them their product (or in the latter two cases sometimes only the idea of the product), they can count on an interpersonal multiplier effect.

## Public health

The limitations of the broadcast approach and research on opinion leaders turned the attention of public health officials to new models of disseminating health information by trying to exploit interpersonal influence. During the 1980s, health promotion programs were increasingly built on community-based intervention that tried to differentiate populations with respect to their health-related behavior (Shea and Basch, 1990). This approach drew upon the theoretical perspective of diffusion (Rogers, 2003) and social learning theory (Bandura, 1977), arguing that people acquire new behavior from people in their environment through observational learning. One well-known example of this approach is the North Karelia Project on smoking. Launched in 1972 in Finland to combat the country's record high mortality of cardiovascular diseases (Puska and Uutela, 2000), this project was one of the first major community-based projects for cardiovascular disease prevention. It was built in partnership with WHO and targeted communities with health information through various channels (television, newspapers, personal health training, seminars, etc.)

in an attempt to reduce the number of smokers. Based on the two-step theory of the diffusion of innovations, a network of local opinion leaders was identified in each community often through relevant local organizations. Opinion leaders were then trained to spread the advisory in a credible and effective manner. Unlike broadcasting, community intervention (as its name suggests) aims at a much smaller audience and assumes a simple, two-step connectedness among audience members.

### 2.2.3 Multi-step model

The concepts of the opinion leader, early adopter, and market maven and the strategic actor of the community intervention approach drew attention to the fact that information cannot be thought of as a one-step process and that the public is not atomized but linked through social ties. Yet all of them, with the exception of the early adopter, lead only to a two-step model. Society or the community is neatly divided between leaders and followers, everyone is either the former or the latter but not both. Studies of early adopters, on the other hand, opened up the possibility of a multi-step diffusion model. As a few, very adventurous early adopters are followed by less adventurous early adopters, and then by mainstream adopters, and finally laggards, the process leads to a chain of diffusion where people in the middle are both leaders and followers. Empirical studies of diffusion began in the late 1930s and were mostly concerned with the spread of innovation, such as the adoption of ham radios (Bowers, 1937), progressivist policies (McVoy, 1940), hybrid corn (Ryan and Gross, 1943), and the prescription of new drugs by doctors (Coleman *et al.*, 1957) through imitation. Most studies were interested in measuring the accumulation of adoptions over time.

In a broadcast model, the curve plotting the cumulative number or proportion of adopters as the broadcast is repeated again and again is close to a logarithmic function, truncated at or before the point where no person is left in the population to adopt. The marginal return to broadcasting is highest at the earliest part of the process. The first broadcasts have the greatest effect and subsequent ones produce a declining yield. Once we move away from the broadcast model and allow for imitation, the typical diffusion model becomes an ogive in the shape of an S, as the diffusion process takes off slowly then accelerates when adopters achieve a critical mass or tipping point (Schelling, 1978; Galdwell, 2000). The curve would slow down as the pool of potential adopters begins to shrink, completing the figure. There are several functions that can describe the empirical curve and much of diffusion research consists of correlating curves with the diffusion mechanisms that are thought to have produced them (Mahajan and Peterson, 1985). Until recently, most diffusion models have been macro models predicting only the total number of adopters. As they rarely observe the transmissions, diffusion researchers mostly deduce how they happen from the aggregate outcome by assuming simple transmission rules. The simplest models assume a homogenous population where individuals have different propensities to adopt. Models of the adoption chain start from the assumption that there are personal characteristics that make some people early adopters. Personal attributes determine at which stage of the diffusion process people enter, i.e. how many adopters they must perceive to make the move. In information diffusion processes, the most curious and cognitively astute people would be the first to find out about an advisory, others would not pick up the information unless they heard it from a friend, yet others would need to hear it from many to understand and believe it, etc. Here the spread of information would depend simply on how many people there are of each cognitive type. If there are too few people at the beginning of the chain, they will transmit the information to too few people in the second group, not enough to make anyone care in the third group and the diffusion fizzles out early. If there are more people in the first group, but very few in the second group, the outcome could be similar, etc. The thing to notice is that in these models, propensities are inherent to each individual that matters (Granovetter, 1978; Kuran, 1987) after that only the aggregate number – or proportion – of people who already adopted makes a difference. A different approach assumes that all people have the same propensity, but they are located differently in a social structure because they are connected differently to others. Whether to adopt or transmit then depends on their relative position – their spatial (Ryan and Gross, 1943; Schelling, 1978) or social proximity (Coleman *et al.*, 1957) to those who have already adopted.

In recent years, researchers have moved further away from the broadcast model that assumes that the world can be sharply divided into a source and atomized targets, beyond the two-step model of center, leaders, and followers, and past the simplistic, multi-step diffusion models and entered a world where all actors are (1) both potential sources and targets, (2) linked in a patterned network of social ties that makes them more or less likely to transmit information and that influences the overall travel of the information in question, and (3) heterogeneous in terms of the properties that make them more or less likely to send or receive information. In this chapter, we will use percolation models which combine graph theory that captures the structural characteristics of networks with dynamic processes.

### 2.2.4 Structural characteristics of graphs

In order to describe a graph, we distinguish vertices (in our case, people) and edges (their ties). To characterize a graph, it is useful to compute certain properties and we present several useful ones for social networks (for a survey, see Goldenberg *et al.*, 2001). Perhaps the most well known property of a social network is that they are often "small worlds," which means that the shortest path length between two people is often small (Milgram, 1967). A formalization of this measure is the mean geodesic distance for a connected graph, which is the average of the shortest paths between all possible pairs of vertices. Another useful measure for social networks is the clustering coefficient of a graph, which is the chance that a friend of your friend is also your friend.

### 2.2.5 Random graphs

Random graphs are formed by generating a set of edges for a graph in a random fashion. Many random graph models have been extensively studied, both theoretically and in relation to real networks, and later we apply them in our simulated graph processes.

A classic random graph model is that of Erdős and Rényi (Erdős and Rényi, 1959, 1960), which was independently discovered by Solomonoff and Rapoport (Solomonoff and Rapoport, 1951). In the model, for a graph of size $n$, each pair of vertices has an edge between them with probability $p_e$. One of the properties of the Erdős–Rényi random graph in the context of social networks is that the clustering coefficient is often quite small. The SW model tends to have a high clustering coefficient and small geodesic distance. It can be constructed by taking a one-dimensional lattice of $n$ vertices in a ring, and joining each vertex to its neighbors $k$ spaces away on the lattice. There are therefore $kn$ edges in this lattice. The edges are re-wired by going through each one in turn and, with probability $p_s$, moving one end of the edge to a new vertex chosen uniformly at random. No self edges (edges from a vertex to itself) or double edges (pairs of edges between the same pair of vertices) are created.

An interesting question about random graphs is how they compare to real-world ones. Dekker (2007) finds that real social networks have a low average network distance, a moderate clustering coefficient, and an approximate power-law distribution of node degrees. In Newman *et al.* (2002) the authors observe that real social networks have highly skewed degree distributions which can vary according to the property being measured. A random graph model is proposed which can be fitted to an arbitrary degree distribution. Such a model is useful since the theoretical computations of clustering and path length over the model often, but not always, bear strong similarities to those found on real data.

### 2.2.6 Modeling dynamic processes
#### *Disease spread: epidemiological models*

In epidemiology, there are two deterministic models used most often to study the spread of infectious disease from person to person. These models are based on a simple mathematical formulation that does not take into account network properties (Hethcote, 2000). The SIR model (people are either Susceptible, Infected, or Recovered) is an appropriate approximation for diseases that infect a significant part of the population in a short outbreak (such as influenza). This model considers people who recovered from the disease to have acquired permanent immunity.

In the SIS (Susceptible/Infected/Susceptible) model, people do not acquire permanent immunity and return to the state of susceptibility when they recover from a disease (e.g. tuberculosis). The SIS model is appropriate for endemic diseases which persist in a population for many years.

Even though the models now being applied to specific diseases are more complex and refined (Hethcote, 1989) – for instance some models use periodic contact rates to take into account the prevalence of many diseases which varies because of seasonal changes in daily contact rates – they are quite limited. One of the limits of these basic models is that they make the unrealistic assumption that the population is homogeneous or "fully mixed." The homogeneity assumption, which states that everyone is equally susceptible before infection and after acquiring the sickness, can be relaxed somewhat assuming that the population belongs to a small set of categories (men, women, or adults/children) with heterogeneous characteristics.

These models also assume that the population is equally and randomly connected (Watts, 2003; Brauer, 2005). Different network structures are addressed by positing multiple levels of mixing, e.g. people may belong to two levels: to a household and to the world, and connected more at one level (with household members) than at another (everyone else, see Ball *et al.*, 1997), yet those solutions still miss many dimensions and configurations of social ties.

*Information spread*

Most empirical studies of information spread investigate diffusion of information through the Internet. For instance, information propagation in Weblogs or "blogs" is analyzed using a corpus containing 401,012 posts in Gruhl *et al.* (2004). One of the dimensions of analysis is the topics of posts. They characterize topics as: "just spike," which are inactive, then active, and then inactive again; "spikey chatter", which have a significant chatter level and are sensitive to real world events and hence have spikes; and "mostly chatter," which have moderate levels of discussion.

The authors also model topic propagation amongst individuals. An information propagation model is derived based on the Independent Cascade model (Goldenberg *et al.*, 2001). In this extended model, each vertex is a person and each directed edge has a probability of information being copied from one vertex to another in the next time quantum. The model is extended with an additional edge parameter, which is the probability that a person reads another person's blog. Edge probabilities for the transmission graph are learnt using an EM-like algorithm (Dempster *et al.*,

1977). One of the observations made from the learnt transmission graph is that most people transmit on average to less than one additional person, whereas some users transmit to many others, providing a boost to certain topics. An important difference between this work and ours is that individuals can reach many others through their blog posts, and in our case, the number of possible transmissions is limited by the number of regular social contacts.

Kempe *et al.* (2003) consider the problem of selecting the most influential nodes in a social network in the context of information propagation. One application of this work is in the analysis of the word of mouth effect in the promotion of products. The challenge is to discover which individuals should be targeted with information in order to trigger a cascade of further adoption. This problem is NP-Hard, however efficient greedy algorithms are shown to find a solution within sixty-three percent of the optimal for the Independent Cascade and Linear Threshold models. Our diffusion model is essentially a deterministic variant of the information cascade model, and hence this result can also be applied to our work.

*Percolation process*

A percolation process is one in which vertices (sites) or edges (bonds) on a graph are randomly designated either "occupied" or "unoccupied" and one asks about various properties of the resulting patterns of vertices (Newman, 2003). Percolation theory is mainly developed in physics but, as Newman (2003) reminds us, one of the initial motivations of its development in the 1950s was the modeling of the spread of disease, and it is still used in epidemiological studies (Sander *et al.*, 2002). Newman and Watts (1999) used site percolation on SW graphs as a model of the spread of information or a disease in social networks, and Allard *et al.* (2009) deployed a bond percolation model taking into account heterogeneity in the edge occupation probability through a multitype network approach (see also Cohen *et al.*, 2000a). Callaway *et al.* (2000) apply a more general approach that considers the probability of the occupation of a vertex given its degree, $k$.

Our chapter proposes a model with varied susceptibility to infection and shows that, under strong heterogeneity in susceptibility, there are patches of uninfected but susceptible people. The model uses both bond and site percolation in which vertices either have a specific piece of information or do not. The question in this chapter relates to how individuals select among their network of people those interested in the information, and how this information is then diffused to the broader network.

## 2.3 Data and method

There are two approaches to network based research. The first looks at ego networks and takes a sample of individuals, and through a series of questions tries to map out their social ties (Marsden, 2005). These are then analyzed together with other individual attributes in statistical models, assuming independent and random observations. The advantage of ego network research is that it allows for large and representative samples and consequently provides generalizable findings. It also permits the collection of a large amount of information about individuals (egos). This tradition generates data that are strong on vertices, but weak and incomplete on edges. The second approach takes entire populations and maps the relationships among all vertices. The relationships or edges are usually complete and they are observed as opposed to reported, but the data about the vertices tend to be limited. The choice of populations – and therefore the topics – is opportunistic.

Our approach belongs to the first line of investigation but with a unique design. We began with an ego network method, then we followed the network path through which the information was dispersed collecting data on dyads in some cases from both the sender and the recipient of the information at the two ends of the tie.

### 2.3.1 Data

To recruit a sample representative for gender, age, and socio-economic status of the French population, we first broadly sent an invitation to 24,000 people to answer a questionnaire on "Food Habits and Food Risks." They answered socio-demographic questions which allowed us to select those who fit the quotas. Although the sample is broadly representative for the French population, people in the sample have more frequent connections to the Internet than the population as a whole. In the sample, all people use an Internet connection, whereas 39% of the French never use one (Cardona and Lacroix, 2008) and 84.7% of people in the sample use an Internet connection at least once a day, while this number is 41% nationwide.

The survey was conducted in three waves. The first wave, which took place in December 2008 and January 2009 interviewed 6,346 individuals, who we call Egos. Egos are those who receive the information from us. In our models, Egos can only be senders of the information as they are at the root of the diffusion process. All interviews were conducted through the Web using self-administered Web surveys. In the first wave, we asked Egos a series of questions about their knowledge of food risks, their food habits, social networks, and socio-demographic characteristics. We also exposed them to the information about Campylobacter, which explained the health hazard and provided advice on how to avoid it. We did not tell them that we were interested in information diffusion, but we informed them that we would revisit them for a follow up.

We returned to the sample of Egos three weeks later. In the second wave, we were able to interview 4,496 of the first wave sample, an attrition of twenty-nine percent. The analysis shows that those who dropped out tended to be somewhat

Table 2.1. *The type of contact through which the information was transmitted.*

| Type of contact | Percent |
| --- | --- |
| Face to face | 78.95 |
| Phone | 15.37 |
| Internet (e-mail, Skype, etc.) | 5.68 |
| Total | 100 (N = 7,655) |

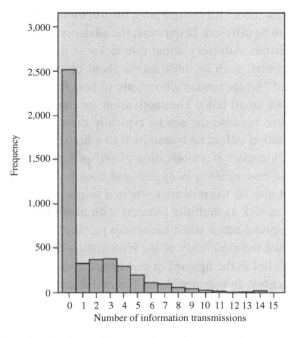

Figure 2.1. The distribution of the number of transmissions reported by Egos (mean = 1.700; Std. Dev. = 2.591; N = 4,496).

less educated. This time we asked people what they remembered of the information about Campylobacter, how they changed their behavior, and if they told about Campylobacter to anyone else. We asked a series of questions about each person they reported to have talked to about Campylobacter, and we requested that they write in their contact's e-mail address and send an email requesting that they fill out a questionnaire for the study. Egos contacted and described 7,655 contacts, which we call Alters. On average, Egos transmitted the information to 1.7 Alters (Figure 2.1). The most common way to convey the information was face to face followed by phone conversations. Only less than 6% of Egos told others about Campylobacter via the Internet (Table 2.1). None of the socio-demographic variables were strongly correlated to transmission but young people, people working in small companies, and those who live with a partner and have children, and those with more than elementary education were a little more likely to transmit the information.

In the third wave, we interviewed those Alters willing to respond to an Ego's request. The questionnaire for these Alters was similar to the first and second questionnaires administered to Egos, adjusting for differences in context.[1] We obtained 451 responses. Only 301 had any recollection of the encounter with an Ego. We used only these Alters in our analysis. The responding Alters reported to have contacted yet another 138 people with the information.

There is no generic model for the spread of information independent of the nature of the information to be diffused. In our case, the advisory about the Campylobacter has certain peculiarities. Advisory about this sickness is not knowledge typically sought by its recipients, such as information about chronic diseases or jobs, but rather, it is "pushed" by its sender who wants to benefit others (or simply wants to pass the time with small talk). The motivation for transmitting the information is rarely self-interest, because the sender typically does not directly benefit from sending the information unless he transmits it to a household member who cooks for him, in which case there is a motivation of self-protection against the disease.

Furthermore, the transmission is dyadic and does not involve a critical mass. In this respect, it is similar to epidemics where a single contact with an infectious person makes one as sick as multiple contacts with many ill people. Certain types of information do spread better when the sender receives it from multiple sources. This is the case when the credibility of the information depends primarily on how widely it is held. Belief in the upward or downward trajectory of the stock market belongs in that category. In our model, the Campylobacter advisory is transmitted dyadically partly because its content is found to be credible by most people (the average score for both Egos and Alters was 6.1 on a seven-point scale) and there is no need for outside reinforcement.

Moreover, health advisories, like most diseases, are most infectious soon after they are received and as time passes they become less likely to be transmitted. In our data, 91% of Egos who could recall the timing transmitted the information during the first week and almost half of those on the day they received the information from our survey. Only less than a tenth of the transmissions happened in the second and third weeks. During the time window of our study we observed most of the transmissions that were going to happen and we are unlikely to have missed an outburst of diffusion after we had completed our study.

---

[1] For instance, we could not ask any questions of Alters that would gauge their knowledge about Campylobacter before they received the information from an Ego. We did ask such questions of Egos in the first wave before we presented the advisory.

## 2.4 Computational simulation

There are several limitations of collecting real data, influenced by factors such as cost and privacy. To alleviate these limitations we extract patterns from the survey data and then extrapolate them by simulating a social network. The main aim of this simulation is to measure the average information spread from each Ego on different types of random network. Average information spread is computed using a measure called *average hop length*, which is defined as the the total number of transmissions of information divided by the total number of original senders.

All computational code is implemented using Python and the NumPy, SciPy, Matplotlib, and Mlpy (Albanese *et al.*, 2009) libraries. The Support Vector Machine (SVM) code is provided using LIBSVM (Chang and Lin, 2001).[2]

### 2.4.1 Data preparation

In the simulation, vertices represent people and edges are relationships between them. For example, an undirected edge exists between vertices if they know each other as friends, family members, colleagues, or acquaintances. Vertices are $\mathbf{v} \in \mathbb{R}^d$, $d = 62$, and store values such as the age, gender, education level of each person. Appendix A shows the complete list of fields used in the simulation. Note that missing values are often replaced with the mode. Furthermore Q7, Q43M1-10, Q47CM1-5, and Q47EM1-5, are categorical variables and hence represented using binary indicator features. As an example, Q7 represents the respondent's profession and has eight categories, hence it is represented by eight binary variables indicating the presence or absence of each category.

The full details of the data preparation stage are given in Appendix B. Essentially, missing characteristics in the data are completed based on the knowledge acquired about the survey population. The total size of the dataset after the completion step is 86,755 pairs of people (examples), in which 82,485 were negative (no transmission) and 4,270 were positive (transmission) examples. We denote by $S$ the set of triples composed of pairs of people and an indicator label for transmission occurrence, $S = \{(\mathbf{v}^{(1)}, \mathbf{v}^{(2)}, y) : \mathbf{v} \in (E \cup A), y \in \{-1, +1\}\}$, where $y = +1$ indicates information transmission from $\mathbf{v}^{(1)}$ to $\mathbf{v}^{(2)}$.

### 2.4.2 Learning transmissions

Given $S$ one must first find a function such that $f'(\mathbf{v}^{(1)}, \mathbf{v}^{(2)}) \approx y$, and we use an SVM (Boser *et al.*, 1992) to find this function. To simplify notation let $\mathbf{x} = (\mathbf{v}^{(1)'} \mathbf{v}^{(2)'})' \in \mathbb{R}^{114}$, and the function mapping $\mathbf{x}$ to $y$ be $f$. We will refer to $\mathbf{x}$ as

---

[2] The complete source code for the simulation experiments is available online at http://sourceforge.net/projects/apythongraphlib/.

an *example* and $y$ as the corresponding *label*. An SVM finds a hyperplane which separates the set of examples into those which are positively labeled and those which are negatively labeled. It does this with maximal margin, which often ensures good generalization to unseen data. A further advantage of SVMs is that they can operate in a kernel-defined feature space and hence model non-linear functions without explicit computation of the new feature space.[3]

As a first step to learning transmissions, the complete dataset is standardized so that the examples have zero mean and unit standard deviation. They are then randomly sampled into a subset of size 15,000. This subset is used for choosing the SVM penalty parameter $C$, and the kernel parameters. For this model selection stage, we use a linear kernel with parameter values selected as $C \in \{2^2, 2^3, \ldots, 2^{11}, 2^{12}\}$, and also the Radial Basis Function (RBF) kernel with $C \in \{2^2, 2^4, \ldots, 2^{10}, 2^{12}\}$. The RBF kernel is given by $\kappa(\mathbf{x}, \mathbf{z}) = \exp(-\|\mathbf{x} - \mathbf{z}\|^2 / 2\sigma^2)$ with kernel width $\sigma \in \{2^{-4}, 2^{-3}, \ldots, 2^1\}$ in this case. As the dataset has many more negative examples than positive ones, the penalty on the errors on positive examples is weighted according to $C^- = C\gamma$ where $\gamma \in \{2^2, 2^3, 2^4\}$.

In order to choose a set of parameters we use $k$-fold Cross Validation (CV) to evaluate prediction error. In this procedure, the dataset is split into $k$ equal sets. One set is kept back for evaluating error, and the remaining are used for training. This is repeated a total of $k$ times with a different test set used each time, and this whole process is repeated for each unique set of SVM parameters. In our case $k = 3$. The output at the model selection phase is the set of parameters for the SVM which result in the lowest error.

Following model selection, one would like to obtain an unbiased estimate of the error obtained on an unseen set of examples. The model selection phase resulted in the selection of the RBF kernel, $C = 2,048$, $\sigma = 1$ and $\gamma = 32$. We use the set of 71,755 examples disjoint from that used during model selection, and a five-fold cross validation procedure to obtain an average error with the SVM using this set of parameters. The resulting balanced error is 0.092 (0.000)[4] which compares favorably to the error obtained on predicting no transmissions, which is 0.5. The error on the positive examples is 0.111 (0.000) and the error on the negative ones is 0.073 (0.000). With the linear kernel, $C = 128.0$ and $\gamma = 16$, the best balanced error rate is 0.235 (0.000) with errors of 0.276 (0.000) and 0.195 (0.000) on positive and negative examples.

The model weights describe the net influence of the various factors determining whether the information is transmitted from Ego/Sender to Alter/Recipient (Table 2.2). These weights are derived from a model that predicts the pattern of

---

[3] See Shawe-Taylor and Cristianini (2004) for an overview of kernel methods.
[4] The value in parentheses is the standard deviation of the error.

Table 2.2. *Variables with the highest weights for predicting transmission.*

| Variable | | Meaning | Finding transmission is more likely if... | Weight |
|---|---|---|---|---|
| Ego's characteristics | | | | |
| Perception | | | | |
| 1 | Q17bisA#X | Risk perception of Campylobacter (after info) | Ego considers Campylobacter infection risky | 76.002 |
| Knowledge | | | | |
| 2 | Q16#X | Seeking of info (after info) | Ego looked for extra information | −56.268 |
| Social network variables | | | | |
| 3 | Q44BX | Number of weekly direct contacts with colleagues | Ego has fewer direct contacts with colleagues | −64.349 |
| Alter's Characteristics | | | | |
| Demographic variables | | | | |
| 4 | Q185$X | Age | Alter is younger | −51.404 |
| 5 | Q184$ | Gender | Alter is female | −108.127 |
| 6 | Q186$ | Education | Alter has less education | −60.097 |
| 7 | Q187$_2 | Profession: enterprise head, artisan, merchant | Alter is not enterprise head, artisan, merchant | −46.412 |
| Social network variables | | | | |
| 8 | Q178C$ | Frequency of contact with colleagues | Alter spends more time with colleagues | −61.636 |
| 9 | Q178B$ | Frequency of contact with neighbors | Alter spends more time with neighbors | −57.886 |
| 10 | Q179M$_3 | Group membership parental assoc | Alter is not a member of parental associations | −51.585 |
| 11 | Q180C$X | Number of weekly direct contacts with family members | Alter contacts many family members | 46.635 |
| 12 | Q183EM$_2 | Tie with discussion about food risk | Alter talks to household member about food risk | 75.639 |
| 13 | Q196$X | Frequency of Internet connection | Alter uses the Internet less frequently | 125.049 |
| Experience | | | | |
| 14 | Q4A$ | Personally cook | Alter does not cook often | −93.724 |
| 15 | Q24$X | How often eats chicken (fowl) | Alter eats poultry less often | 48.795 |

Table 2.2 *(cont.)*

| Variable | | Meaning | Finding transmission is more likely if… | Weight |
|---|---|---|---|---|
| Knowledge | | | | |
| 16 | Q1A$ | General knowledge about food risk | Alter thinks s/he knows more about food risk | 94.683 |
| 17 | Q21$ | Previous knowledge about Campylobacter | Alter had no previous knowledge about Campylobacter | 94.127 |
| 18 | Q32$X | Seeking of info (after info) | Alter did look for more info on Campylobacter | −63.833 |
| Perception | | | | |
| 19 | Q33A$X | Risk perception of Campylobacter (after info) | Alter considers less risky Campylobacter | −86.062 |
| 20 | Q20A$ | Finds info credible | Alter finds info credible | 49.843 |

transmission with error. As smaller magnitude weights can be due to prediction error, we concentrate on the large weights and cut out the smaller ones where the relative size of the noise is larger. One of our findings is that in our model, the characteristics of Alters matter more than the characteristics of Egos, as seventeen of the twenty variables with the highest weights belong to Alters. This is surprising because it is the Ego/Sender who decides whether to relay the information. There are two explanations for this finding, one is technical the other is substantive. The technical explanation starts from the recognition that we have considerably fewer fully observed Alters than Egos. The characteristics of Alters are predicted and not actually observed. Because we generated the characteristics of recipient and non-recipient Alters differently, and non-transmitting ties always involve on one end a non-recipient Alter and transmitting ties a recipient one, this could have amplified the influence of Alter characteristics on whether or not a tie transmits the advisory.

The substantive explanation, on the other hand, rests on the assumption that the Egos decide to relay the information on the basis of the characteristics of Alters, although, as we will see not by estimating the recipient's need for the message, which they may not know, but on the basis of their perception of the recipient's general interest in food risk. Another mechanism that can explain the importance of Alters' characteristics is that Recipients elicit the advisory. For instance, women raised to be more attentive to others will be more likely to receive the information by the way they communicate with Egos. Because the technical reasons do play some part in the elevated importance of Recipient characteristics, the extent of which we cannot tell, we make our substantive claim with caution.

The factors can be sorted into five categories: demographics, social networks, experience, knowledge, and perception. To interpret the effect of social networks on transmission, we have to keep in mind that our unit of analysis is the social tie and not the individual. The three variables with high weights that describe Ego belong to the perception, knowledge, and network categories and our model does not point to any demographic or experiential variables on the Ego side. Two of the three factors that describe transmitting Egos are not surprising. Egos who perceive Campylobacter more risky and those who looked up additional information are more likely to send the information to Alters. The third one, that the frequency of direct contacts with colleagues has a negative effect, will be explained below.

A tie is more likely to transmit the information about Campylobacter if the Alter is female, young, has less formal education, and if she is not a self-employed entrepreneur by profession.

Of the social network characteristics of Alters, spending more time with neighbors, family members, and colleagues increases the chances of a tie to transmit. Interestingly, frequent contacts with colleagues by Ego have the opposite effect. Ties are more likely to transmit if Ego has fewer contacts with, while Alter spends more time with colleagues. This, however, is less of a paradox than it seems.

We observe that this apparent paradox involves two different aspects of collegial ties: quantity and intensity. The second part of this seeming contradiction, that intensity of ties has a positive effect on transmission, is not unexpected. People who spend more time together with colleagues will be more likely to hear about topics unrelated to work. We find the same relationship for neighbors.

The second part requires more explanation that involves our unit of analysis: ties and not people. Egos who are well connected to colleagues in our dataset will contribute many collegial ties. This means that if Ego has many collegial ties but only a few will carry the information, most ties with collegially well-connected Egos will not carry the information. When we consider ties rather than people, what matters is not whether the number of transmissions rises with the number of ties but whether it rises at a higher rate than ties do. If the non-transmitting ties rise faster than transmitting ones, the relationship for ties between transmission and the number of ties will be positive for people (the more ties the more transmission), but we will observe a negative relationship for ties (the more ties the less likely a tie will transmit). Collegial ties thus have a diminishing marginal return: the first few colleagues will raise the chances of Ego's telling some of them about the advisory, but working with ten as opposed to twenty colleagues makes little difference.

Why is it then that the number of family contacts increases and not decreases the probability of transmission? The answer is that people have fewer familial than collegial contacts. It seems that the general relationship between number of ties and transmission is such that for the first few ties, not just the number but also the rate of transmission rises. Ego must have a certain number of ties to find one Recipient who is interested, the very first or second tie may not make much of a difference, the third and fourth do. Therefore, what we learn from our model is that higher intensity of collegial and neighborly ties of the recipient will make it more likely that she or he receives the information. The number of ties will yield an increasing return in the beginning, but a decreasing one after a certain point.

We also included measures of formal affiliation to organizations as a measure of social networks. We learn that transmission is less likely to be targeted at people who are members of parent associations, and other formal associations do not seem to make a difference.

An indirect measure of a type of connectedness is the frequency with which people use the Internet. Our survey was conducted through the Web, thus it seems strange that people who spend more time on the Web are not more but are less likely to receive the advisory. Yet this result is consistent with our finding that most of the transmission happens face to face (see Table 2.1).

Experience and practices of Alters seem to indicate that transmission is not driven by the need of the recipient to know about Campylobacter. Those who cook more and eat chicken more often are less likely to receive the information about the disease. One possible explanation is that Ego assumes that people who cook a lot and eat lots of chicken are more likely to already be aware of the advisory. What seems to attract the information is if Alter has a general interest in food risk. Ties to recipients who report that they have more general knowledge and then seek additional information on Campylobacter once they received the advisory, i.e. ties to people who are alert and curious about the topic are the most likely to carry the transmission.

As for Alters' perceptions, information seems to decrease their fears as the advisory communicates the measures that can be taken to avoid the infection. This creates a break on the transmission: if perceiving riskiness increases the urge to send the advisory but receiving the information reduces the sense of risk, at the next step, there will be fewer people relaying the information.

### 2.4.3 Modeling diffusion

The SVM model which predicts information transmission is applied to a set of random graphs in this stage. We artificially create a set of vertices and use these vertices as the basis of a graph, and we select a random subset of the vertices to have information about Campylobacter. Using the SVM model, and various initial

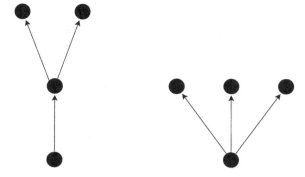

Figure 2.2. An illustration of the difference between $\mu_h$ and $\xi$ measures, in which edges represent information transmissions between vertices. For the graph on the left there is one original sender, two senders, and three transmissions and receivers, hence $\mu_h = 3$ and $\xi = 3/2$. For the one on the right there is only one sender and $\mu_h = 3$ and $\xi = 3$.

setups, we observe how information is diffused within the graph (see Algorithm 1 of Appendix C).

At the end of the algorithm, several measurements made during the simulation are output. The quantities $\nu_1, \ldots, \nu_m$ are the proportions of people with information at each iteration. Another quantity of interest is $\xi$, which is the total number of receivers/total number of unique senders. This is a measure of the "fan-out" of information, i.e. how many vertices receive information directly by each sender, on average (see Figure 2.2).

The simulation is run using 10,000 vertices, with an SVM trained using 20,000 random examples from the generated transmissions. We use the SW model to generate vertices with neighbors $k \in \{10, 15\}$, re-wiring probability $p_s \in \{0.01, 0.05, 0.1, 0.2\}$ and initial information probability $a \in \{0.1, 0.2, 0.5\}$. Algorithm 1 is run five times for each set of parameters and the results are shown in Table 2.3. The same test is repeated with the Erdős–Rényi random graph with $p_e \in \{0.001, 0.002, 0.003, 0.004\}$.

The results with the Erdős–Rényi graphs are shown in Table 2.4. In general, we observe that the number of new vertices receiving information decreases at each iteration, and one would expect this decrease as the probability that neighbors also have information increases with $i$. Since $p_e$ is the probability of an edge, the mean number of edges is expected to be $np_e$, and $\mu_h$ should increase with $p_e$. Table 2.4 shows that this is the case, however a doubling in the probability of an edge from 0.1 to 0.2 results in less than double the average hop distance. Note that the more neighbors a vertex has, the more likely that a greater number of

Table 2.3. *Results from the information diffusion simulation using the SW model. Standard deviations shown in parentheses. The re-wiring probability is $p_s$ and $k$ is the initial number of neighbors. The initial information probability is $a$, $\mu_h$ is the average hop distance, and $\xi$ is the total number of receivers/total number of unique senders. The proportion of vertices with information is recorded as $v_0, \ldots, v_3$.*

| $p_s$ | $k$ | $a$ | $\mu_h$ | $\xi$ | $v_1 - v_0$ | $v_2 - v_1$ | $v_3 - v_2$ |
|---|---|---|---|---|---|---|---|
| 0.01 | 10 | 0.1 | 3.544(.124) | 1.630(.065) | 0.106(.005) | 0.040(.006) | 0.013(.006) |
| 0.01 | 10 | 0.2 | 2.677(.069) | 1.196(.030) | 0.135(.010) | 0.032(.014) | 0.006(.015) |
| 0.01 | 10 | 0.5 | 1.835(.031) | 0.528(.017) | 0.124(.016) | 0.008(.016) | 0.000(.015) |
| 0.01 | 15 | 0.1 | 4.499(.075) | 1.771(.050) | 0.127(.009) | 0.060(.013) | 0.015(.012) |
| 0.01 | 15 | 0.2 | 3.363(.060) | 1.249(.038) | 0.164(.007) | 0.041(.008) | 0.006(.009) |
| 0.01 | 15 | 0.5 | 2.229(.041) | 0.493(.016) | 0.136(.006) | 0.011(.007) | 0.001(.006) |
| 0.05 | 10 | 0.1 | 3.609(.097) | 1.615(.039) | 0.104(.009) | 0.043(.011) | 0.012(.012) |
| 0.05 | 10 | 0.2 | 2.676(.036) | 1.188(.026) | 0.136(.004) | 0.033(.006) | 0.007(.006) |
| 0.05 | 10 | 0.5 | 1.815(.034) | 0.526(.027) | 0.119(.005) | 0.010(.006) | 0.001(.006) |
| 0.05 | 15 | 0.1 | 4.782(.115) | 1.809(.031) | 0.131(.008) | 0.063(.012) | 0.017(.014) |
| 0.05 | 15 | 0.2 | 3.343(.053) | 1.222(.019) | 0.157(.004) | 0.043(.006) | 0.007(.009) |
| 0.05 | 15 | 0.5 | 2.242(.031) | 0.497(.011) | 0.137(.003) | 0.011(.005) | 0.000(.006) |
| 0.1 | 10 | 0.1 | 3.681(.033) | 1.577(.021) | 0.100(.006) | 0.044(.008) | 0.016(.007) |
| 0.1 | 10 | 0.2 | 2.713(.088) | 1.196(.039) | 0.136(.007) | 0.036(.007) | 0.007(.008) |
| 0.1 | 10 | 0.5 | 1.831(.016) | 0.528(.010) | 0.121(.009) | 0.011(.011) | 0.001(.010) |
| 0.1 | 15 | 0.1 | 4.812(.128) | 1.761(.031) | 0.130(.009) | 0.061(.015) | 0.020(.014) |
| 0.1 | 15 | 0.2 | 3.373(.037) | 1.226(.031) | 0.158(.007) | 0.043(.008) | 0.007(.008) |
| 0.1 | 15 | 0.5 | 2.244(.057) | 0.495(.016) | 0.139(.005) | 0.011(.005) | 0.001(.005) |

Table 2.4. *Results from the information diffusion simulation using the Erdős–Rényi model. The probability of an edge is $p_e$.*

| $p_e$ | $a$ | $\mu_h$ | $\xi$ | $v_1 - v_0$ | $v_2 - v_1$ | $v_3 - v_2$ |
|---|---|---|---|---|---|---|
| 0.001 | 0.1 | 2.900(.098) | 1.394(.015) | 0.063(.011) | 0.031(.015) | 0.012(.015) |
| 0.001 | 0.2 | 2.122(.081) | 1.141(.027) | 0.090(.011) | 0.028(.014) | 0.006(.015) |
| 0.001 | 0.5 | 1.460(.013) | 0.624(.026) | 0.092(.008) | 0.008(.007) | 0.001(.006) |
| 0.002 | 0.1 | 4.112(.056) | 1.560(.016) | 0.099(.005) | 0.056(.010) | 0.022(.012) |
| 0.002 | 0.2 | 2.790(.065) | 1.148(.032) | 0.130(.010) | 0.040(.015) | 0.009(.015) |
| 0.002 | 0.5 | 1.829(.028) | 0.539(.010) | 0.120(.005) | 0.010(.006) | 0.001(.006) |
| 0.003 | 0.1 | 5.130(.167) | 1.678(.036) | 0.129(.010) | 0.073(.018) | 0.022(.020) |
| 0.003 | 0.2 | 3.455(.102) | 1.210(.029) | 0.159(.009) | 0.044(.012) | 0.009(.013) |
| 0.003 | 0.5 | 2.225(.027) | 0.490(.011) | 0.135(.009) | 0.011(.007) | 0.001(.006) |
| 0.004 | 0.1 | 6.029(.179) | 1.765(.034) | 0.141(.009) | 0.078(.016) | 0.022(.018) |
| 0.004 | 0.2 | 4.123(.033) | 1.264(.021) | 0.178(.007) | 0.049(.008) | 0.008(.010) |
| 0.004 | 0.5 | 2.721(.022) | 0.488(.010) | 0.152(.011) | 0.013(.011) | 0.001(.010) |

those neighbors without information are suitable candidates for transmission (as learnt by the SVM). However, it is clear that as each vertex has more neighbors, the chance of an information-containing vertex coming across a neighbor which also has information increases. In a similar way, an increase in initial information probability corresponds with smaller average hop distances, though a doubling of $a$ results in $\mu_h$ which is greater than half of the original value.

The values of $\xi$ capture a different aspect of the information diffusion. One would expect a higher value of $a$ to imply more neighbors with information for each vertex and hence a lower $\xi$ value, and in general this is the case. When $a = 0.5$, an increase in the value of $p_e$ results in a decrease in $\xi$, possibly since there are more people who are able to send information and fewer who can receive. Notice also that the values of $v_{i+1} - v_i$ have an interesting trend: for $i = 0$ the increase is greatest in most cases when $a = 0.2$, compared to when it is either 0.1 or 0.5

The SW results are given in Table 2.3. Note that the Erdős–Rényi graph is similar to a SW model with a re-wiring probability of 1. Hence, a useful comparison is between the Erdős–Rényi graphs with $p_e = 0.003$ and the SW graphs with $k = 15$ and $p_s = 0.1$. The interesting differences in this case are the values of $\xi$. The values of $\xi$ are smaller in the Erdős–Rényi graphs, and since, surprisingly, the total number of receivers is approximately the same, it implies that there are more unique senders in these graphs. Recall that with high clustering the chance of a friend of a friend being a friend is high, and hence fewer senders are able to receive the same number of people as compared with the random connectivity of the Erdős–Rényi graphs.

Another interesting finding is that at the lowest level of $a (a = 0.1)$ the Erdős–Rényi graph results in a higher total number of Alter recipients than the SW network, but as $a$ increases, the difference disappears. This implies that if the initial broadcasting of the message reaches only a few people, SW networks are less efficient, but if the number of Egos gets above a certain proportion of the population, there is no difference between the two network types in terms of how many Alters the diffusion process delivers.

The overall number of receivers does not vary significantly with changes in $p_s$, however for fixed $p_s$ and $k$, the total number of receivers is highest when $a = 0.2$. This latter trend is not generally observed with the Erdős–Rényi graphs. Clearly, the total number of transmissions rises as more edges can accommodate transmissions and falls as the network becomes saturated. The peak is generally before 0.1 with Erdős–Rényi and approximately 0.2 with SW graphs.

Several of the other trends present in Table 2.3 are trivial. An increase in $k$ from 10 to 15 always results in higher transmission since each vertex has more edges and hence more chance of passing information. The length of information paths is short for high values of $a$, implying that the person who receives information passes it onto many others but those receivers rarely pass it on. Furthermore, when $a = 0.5$ the information often only travels along a path of length 2.

In summary this relates to information propagation in the following ways: network structure does not seem to influence the total number of people receiving information. What is important is the number of connections and the probability of having information in the first place. Clearly, not every pair of people will facilitate a transmission and hence if too few people are provided with information, then it may stop before reaching everyone interested in it. Similarly, a saturated network does not permit a lot of transmissions. In the SW model the maximum total number of receivers occurred when 20% of the population were provided with the information. Since $v_{i+1} - v_i$ always decreases with $i$, this information is clearly in contrast with "viral" information spread in all of the scenarios presented.

### *Transmission visualization*

The measures recorded in Tables 2.3 and 2.4 give a good idea of the information diffusion processes occurring in the generated graphs. We additionally consider the visualization of graph transmissions. We start with the SW model with $n = 10,000$, $k = 15$, $p_s = 0.1$, and a value of initial information probability of $a = 0.1$. In this particular instance, the total number of recipients (including the initial ones) is 2,897. Disregarding orientation of information transmission, 2,346 persons belong to unique connected components, while the remaining 551 persons fall into much smaller components (the largest having thirty-five members). However, in spite of constant progress in graph visualization (Di Battista *et al.*, 1999; Herman *et al.*, 2000), representing the main connected component in a legible form remains impossible. We rely therefore on two simplifying assumptions.

The first one consists of building a clustering of the nodes of the main connected component (Schaeffer, 2007; Fortunato, 2010): we find groups of individuals which are more likely to transmit or receive the information inside their group than to members of other groups. Then, rather than displaying the original large graph, we draw a graph of the clusters: each node corresponds to a group of people from the original graph (the surface of the node is proportional to the number of persons in the cluster). The edges between nodes indicate information transmission between members of the corresponding clusters. Concretely, we use maximal modularity clustering (Newman and Girvan, 2004) with the algorithm described in Noack (2007) and Noack and Rotta (2009), using the implementation provided by Andreas Noack.[5] The visualization of the clustered graph is done using the Fruchterman–Reingold algorithm (Fruchterman and Reingold, 1991) as implemented in the Igraph R package (Csárdi and Nepusz, 2006; R Development Core Team, 2010).

---

[5] Available at http://code.google.com/p/linloglayout.

The clustering process finds fifty clusters: less than 3.7 percent of information transmission happens between clusters while the rest takes place inside clusters. This provides a validity index for the clustering: ignoring information propagation outside of a cluster will not introduce major distortions in the analysis. The resulting display is given by Figure 2.3, upper left panel. The figure also shows information

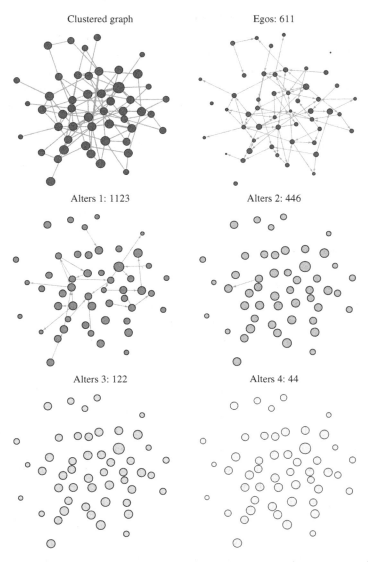

Figure 2.3. A clustered representation of the largest connected component of propagation graph used in Section 2.4.3 (upper left panel) and a clustered representation of information propagation; see main text for details.

propagation in the graph. Apart from the upper left panel, each graph of this figure shows the number of persons that have received the information at each time step of the propagation (the surface of each node encodes the number of persons). The upper right graph corresponds to the initial receivers (611 Egos), while arrows show information propagation from one cluster to another. The left graph of the second line shows the number of receivers after one step of propagation (i.e. Alters 1), etc. The fact that most of the propagation happens between clusters during the first two steps is easily explained by two factors. Firstly, the initial growth is the largest one and corresponds to the largest numbers of transmissions: it should generate the largest part of the external transmissions. Moreover, the clustering algorithm used is based on modularity maximization. Modularity is a quality criterion for graph clustering that rewards putting connected nodes in the same cluster. However, the reward is inversely proportional to the degree of the nodes. Therefore, the obtained clustering tends to put high degree nodes in different clusters. As the first information transmission is the one in which persons tend to pass knowledge to the largest number of alters, those egos are more likely to be assigned to distinct clusters than the transmitters of the following steps.

As most of the information propagation happens inside clusters, focusing on one cluster provides a good idea of the general transmission. We display information propagation in the largest cluster in Figure 2.4. To avoid missing information propagation, the cluster was extended in the following way: when information flows *from* a person *in* the cluster under consideration *to* a person in *another* cluster, the recipient is added to the cluster. In the present case, the cluster grows from an initial size of 96 persons to 112 persons. Then, the result presented in Figure 2.4 is exactly the propagation that would have happened, even if no other persons apart from those in the cluster had initially received the information.

## 2.5  Distribution of individual (vertex) characteristics

In this test we run a simulation using the SW model with parameters $p_s = 0.1$ and $k = 15$, and observe how the distribution of various characteristics such as gender, age, etc. varies at each iteration. Transmissions are learnt using a sample of 20,000 examples, and an SVM is trained using the parameters found in Section 2.4.2. The simulation is run with 20,000 vertices for three iterations and repeated a total of five times with different random seeds. We observe that there are 1,495.2 Egos, and 2,689.6, 1,245.2, 369.4 new Alters iterations 1, 2, and 3, respectively (Figure 2.5).

The simulation shows that after the initial jump the number of receivers tapers down. The gender composition changes with each round of iteration. As we have seen earlier, women are more likely to be the recipients of the information than men. This explains that there is an overall increase in the proportion of women.

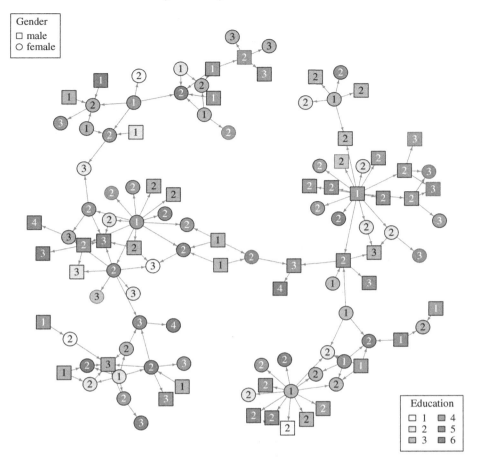

Figure 2.4. Information propagation in the largest cluster, extended as explained in the main text. Education levels are encoded by greyscale and genders by shape. The iteration number at which the information reached the person is written in the corresponding node.

This increase mostly levels off after the first round of transmission. This is due, not just to the fact that there are increasingly fewer new recipients of the advisory, but also that subsequent transmissions are more gender balanced. In the second round the proportion of women actually decreases compared to the first one and levels off in the third, and final round (Figure 2.6).

As the information spreads in our simulation we see a similar pattern for the average age of those who receive the advisory (Figure 2.7). The black line in this chart is almost the mirror image of the previous one. There is an initial drop in age then it changes little.

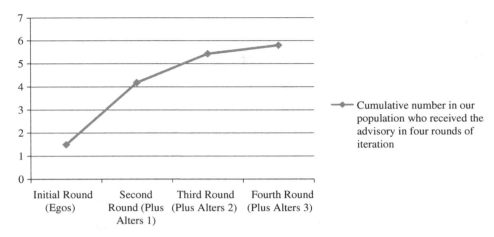

Figure 2.5. Cumulative number in our population who received the advisory in three iterations.

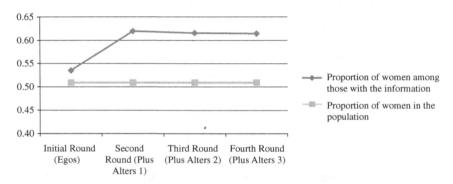

Figure 2.6. The gender distribution in our simulation of four rounds of information diffusion.

We have similar findings for education and general knowledge of food risk (Figures 2.8 and 2.9). Overall the pattern is that the decisive change is in the first round and then smaller changes occur for subsequent rounds, which together with the ever decreasing number of new recipients results in a stable average.

The simulation shows that with the information spreading, the population it reaches is increasingly female, young, less educated and tends to be more knowledgeable about food risks in general, however, much of this shift takes place in the first round of the transmission. This is partly because the characteristics that make people more likely to receive the information are not making them more likely to send it further, and it is also partly because of saturation, as recipients with

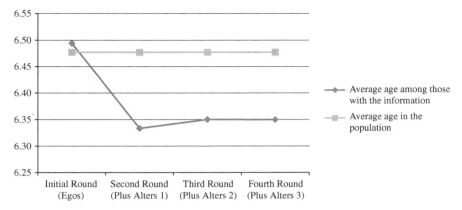

Figure 2.7. The average age in our simulation of four rounds of information diffusion (age is measured on a 12-point ordinal scale: 1 – less than 12, 2 – 12–18, 3 – 19–25, 4 – 26–30, 5 – 31–35, 6 – 36–40, 7 – 41–45, 8 – 46–50, 9 – 51–55, 10 – 56–60, 11 – 61–70, 12 – more than 70).

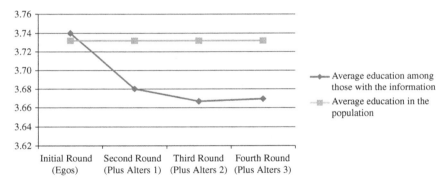

Figure 2.8. The average education in our simulation of four rounds of information diffusion (1 = without degree/primary/BEPC, 2 = CAP/BEP, 3 = BAC, 4 = BAC+2, 5 = more than BAC+2 but no degree, 6 = doctorate).

those characteristics not having the information in a person's social circle become more scarce.

In terms of occupational categories, the distribution shifts in such a way that, among the information recipients we find a smaller portion of retirees and people not working at the end of the third round than we had in the beginning (Figure 2.10). The diffusion process reaches the economically active people more successfully than the inactive ones, but, again, we see the same pattern: the first step is the largest one, and then movement is in the opposite direction but in much smaller steps. This pattern is in line with our earlier findings about the role of collegial ties, the

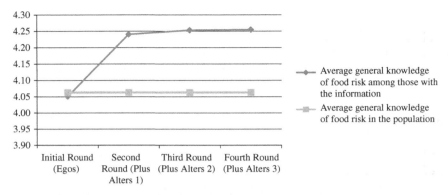

Figure 2.9.  Average general knowledge of food risks on a seven-point scale in our simulation of four rounds of information diffusion.

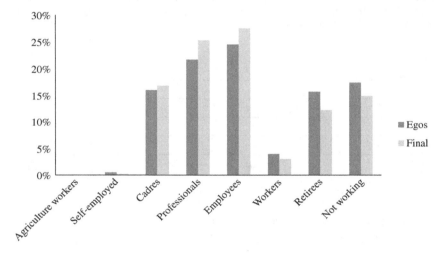

Figure 2.10.  The occupational distribution for Egos at the end of our simulation of information diffusion.

type of ties that only economically active people possess. Because collegial ties are usually within occupational categories, we can explain the seemingly contradictory findings that, net of other factors, Alters with less education are more likely to receive the information but the occupational groups with higher average education, such as professionals and employees, will grow faster than others among those who receive the information. It seems that professionals and employees get the advisory not because they are better educated, but probably because the nature of the workplace interaction they have compared to workers. Within each group, however, it is the less educated who are more likely to be given the news about the Campylobacter.

When we re-run the diffusion using the comparable Erdős–Rényi model, the distribution of the characteristics is very similar and the differences are within the range of random error.

## 2.6 Conclusion

The diffusion of the advisory about the Campylobacter shows that the broadcast model misses an important part of the spread of this type of information: the diffusion of the advisory from its initial audience to others. In contrast to the two-step process model that, in an attempt to identify opinion leaders, emphasizes the characteristics that make people good senders of information, we found that the characteristics that make people good recipients are equally, if not more important. Modeling the diffusion showed us that random networks and SW networks produce roughly the same results in terms of the proportion of the population that receives the information, but the less clustered random networks achieve the result through longer hop distances and a smaller ratio of receivers to unique senders. The composition of those who receive the information changes most in the first transmission, and then it levels off not just because there are fewer and fewer recipients making it harder to change the cumulative total, but also because new recipients in each round are less different from the overall population. This is partly because the characteristics that make someone more likely to receive the information are not those that make them more likely to pass it on. The mismatch between the two sets of characteristics will dampen transmission even in the absence of network barriers.

The diffusion of the Campylobacter advisory shows that word of mouth dispersion of the information, an aspect of the process the broadcast model ignores, is significant. It also demonstrates that while the first transmission from original Egos to Alters is the biggest part of this word of mouth dispersion, the two-step model is flawed because there are still substantial – although smaller – steps spreading the message it fails to recognize. The two-step model also focuses on the qualities of the sender (opinion leader), and our findings show that the recipients' qualities are important. Furthermore, our percolation model was able to build a complex dispersion process which allowed us to account for the heterogeneity of people and of their social networks. Our simulations also showed that not just the local but the global properties of networks matter. Replicating this study with other advisories will tell us the extent to which our empirical findings can be generalized and be of practical use to state agencies.

## Appendix A List of features used in simulation

Table A.1 *The list of features used for each vertex.*

| ID | Field name |
|---|---|
| Q4 | Gender |
| Q5 | Age in categories |
| Q48 | Education |
| Q7 | Profession (8 categories) |
| Q51 | Income |
| Q52 | Single or married |
| Q53 | Number of children |
| Q13 | Size of village/town/city |
| Q20A | Personally cook |
| Q26 | Frequency of eating poultry meat |
| Q28A | General knowledge about food risk |
| Q22 | Personal experience with food risk |
| Q34 | Previous knowledge about Campylobacter |
| Q16# | Seeking of info (after info) |
| Q55 | Frequency of Internet connection |
| Q17bisA# | Risk perception of Campylobacter |
| Q37A, Q37B | Finds the information credible, understandable |
| Q37C, Q37D | Finds the information convincing to change behavior, worrisome |
| Q42A-D | Contact frequency with friends, neighbors, colleagues, family |
| Q43M1-10 | Organization memberships (1 organization per field, 10 options) |
| Q44A-D | Weekly contact with friends, colleagues, family, acquaintances |
| Q45 | Friends' closeness |
| Q47CM1-5 | Tie with discussion about health (6 categories) |
| Q47EM1-5 | Tie with discussion about food risk (6 categories) |
| Q50 | Number of people at physical place of work |
| Q46A-D | Homophily of contacts (age, gender, education, income) |

## Appendix B Data completion

There are $q = 4,496$ Egos, denoted by $E = \{\mathbf{c}_1, \ldots, \mathbf{c}_q\}$ with $c_i \in \mathbb{R}^{62}$ for all $i$. The Egos collectively listed a total of 7,655 people with whom they discussed the Campylobacter information. Among them, there are $r = 301$ Alters who responded to the survey for whom we got the set of 62-dimensional vectors, $A = \{\mathbf{d}_1, \ldots, \mathbf{d}_r\}$. The information recorded for the set of other receivers (7,204 Alters) is limited to their gender, age, profession, and education.[6] To make a prediction for when information transmission occurs between two people, one needs the complete, immediate, social network for Egos, and the 62-dimensional vector for Alters. Hence, we completed the data by generating a set of non-receivers and completing the set of receivers (Alters) for each Ego.

To compute non-receivers, we consider two characteristics: one is the number of times Egos have direct contact with friends, colleagues, family members out of the household and acquaintances in a typical week (Q44A-D). The total number of non-receivers is computed using Q44A-D. The other is whether most of the Egos' contacts are homophilic (i.e. they are similar in their characteristics to the Ego, Q46A-D). Homophily can be related to sex, age, education, and income. The set of homophiles for the $i$th person $H_i \subseteq E$ is the set of people with identical features as indicated by Q46A-D. For example, if a person $c_i$ states in Q46A-D that most of their contacts have the same age and gender, then $H_i$ is the set $H_i = \{c_j : c_j \in E, c_{ik} = c_{jk}, c_{i\ell} = c_{j\ell}, i \neq j\}$ where $c_{ik}$ is the $k$th element of $c_i$, and $k$ and $\ell$ are the indices corresponding to age and gender respectively. Similarly the set of non-homophiles is $H_i' = E \setminus (H_i \cup \{c_i\})$. Given the total number of contacts $N_i$ for the $i$th person and a probability $h$ of being a homophile, a set of $hN_i$ vectors is randomly sampled from $H_i$ and $(1-h)N_i$ vectors are sampled from $H_i'$. In our case $h = 0.7$.

As previously stated, the Egos record only the age, gender, profession, and education of their Alters, and one needs to complete the data using $A$. For the $i$th person in $E$ and the $j$th Alter recorded, we find a random element of the set of Alter homophiles $G_i = \{d_j : d_j \in A, d_{ik} = d_{jk}, d_{i\ell} = d_{j\ell}, d_{im} = d_{jm}\}$ where $d_{ik}$ is the $k$th element of $d_i$, and $k, \ell, m$ are the indices corresponding to age, gender, and education respectively.

## Appendix C Information diffusion algorithm

In Algorithm 1, the vertices of the graph are generated using a multivariate normal distribution, with the mean vector and covariance matrix computed using the Egos and Alters from the survey data. Edges in the graph are added according to the Erdős–Rényi or small-world graph models. At line 1, $a|V|$ vertices are selected at random and marked as having information. The following loop (i.e. the inner "for loop" iteration statement) iterates through all of the edges in the graph, and makes a prediction as to whether information is transmitted along that edge using the characteristics of the vertices and the Support Vector Machine (SVM) model

---

[6] An additional 150 Alters failed to remember the information provided to them.

learnt at line 1. Note that the original survey data are not necessarily identically and independently distributed, but we assume so in order to apply the SVM.

---

**Algorithm 1** Pseudo code for information diffusion simulation.

1: **Input:** Graph size $n$, iterations $m$, initial proportion with information $a$
2: Learn SVM model of information transmission
3: Create $G = (V, E)$ with $|V| = n$ randomly generated vertices
4: Randomly select a set $I_0$ of $a|V|$ vertices to mark with information
5: **for** $i = 1$ to $m$ **do**
6:     **for** $j = 1$ to $|E|$ **do**
7:         Make information transmission prediction along edge $j$
8:     **end for**
9:     $I_i$ is the set of vertices with information, let $v_i = |I_i|/n$
10: **end for**
11: Set $\mu_h$ as total number of transmissions/total number of original senders
12: Set $\xi$ as total number of receivers/total number of unique senders
13: **Output:** Proportions $v_1, \ldots, v_m$ of people with information, $\xi$ and $\mu_h$.

---

# 3

# Scientific teams and networks change the face of knowledge creation[1]

BRIAN UZZI, STEFAN WUCHTY, JARRETT SPIRO, AND
BENJAMIN F. JONES

There is an acclaimed tradition in the history and sociology of science that emphasizes the role of the individual genius in scientific discovery (Merton, 1968; Bowler and Morus, 2005). This tradition focuses on the guiding contributions of solitary authors, such as Newton and Einstein, and can be seen broadly in the tendency to equate great ideas with particular names; for example: the Heisenberg uncertainty principle, Euclidean geometry, Nash equilibrium, and Kantian ethics. The role of individual contributions is also celebrated through science's award-granting institutions, like the Nobel Prize Foundation (English, 2005).

However, several studies have explored the evident shift in science from this individual-based model of scientific advance to a collaborative model. By building on classic work by Harriet Zuckerman and Robert K. Merton, many authors have established a rising propensity for teamwork in samples of several research fields; with some studies going back a century (Collins, 1998; Cronin *et al.*, 2003; Merton, 1973a; Jones, 2005). For example, Derek de Solla Price examined the change in team size in chemistry from 1910 to 1960, forecasting that by 1980 zero percent of the papers would be written by solo authors (de Solla Price, 1963). According to our research, the mean team size for papers written in chemistry had grown to nearly 3.7 contributors by the year 2000. Recently, Adams *et al.* (2005) established that teamwork has been increasing over time across broader sets of fields among the most competitive U.S. research universities. Nevertheless, the breadth and depth of this projected shift in manpower remains indefinite, particularly in fields

---

[1] This chapter is a compilation of parts of two separately published studies. The studies quoted here are: Wuchty, S., Jones, B. F., and Uzzi, B. 2007. The increasing dominance of teams in the production of knowledge. *Science*, **316**, 1036–1039; Guimerà, R., Amaral, L. A. N., Spiro, J., and Uzzi, B. 2005. Team assembly mechanisms determine collaborative network structure and team performance. *Science*, **308**, 697–702.

---

Networks in Social Policy Problems, eds. Balázs Vedres and Marco Scotti. Published by Cambridge University Press © Cambridge University Press 2012.

where the size of experiments and capital investments remain small. This raises the question of whether the projected growth in teams is universal or cloistered in specialized fields.

A shift towards collaborative work also raises new questions of whether teams produce better quality science than individual leaders. Teams may bring greater collective knowledge and effort to a project, but they are also known to experience problems in social networking and coordination that make them perform less well than individuals, even in highly complex tasks (Guimerà *et al.*, 2005; Babchuk *et al.*, 1999; Williams and O'Reilly, 1998).

F. Scott Fitzgerald's observation that "no grand idea was ever born in a conference" (Fitzgerald, 1993), could be called on to illustrate the pre-eminence of the solitary thinker. But subscribing to this viewpoint might lead one to the faulty conclusion that the shift towards collaborative work is a costly phenomenon that promotes low-impact discoveries, and that the highest-impact ideas remain the domain of great minds working alone. However, this is just not the case, as our findings show.

## 3.1 Data

In order to determine how much work is currently being done in teams, as opposed to individuals, we studied 19.9 million research articles in the Institute for Scientific Information (ISI) Web of Science database, and an additional 2.1 million patent records. The Web of Science data cover research publications in science and engineering since 1955, and arts and humanities since 1975. The patent data cover all U.S. registered patents since 1975 (Hall *et al.*, 2001). A collaborative team was defined as having more than one listed author for publications, or, in the case of patents, more than one inventor. Following the ISI classification system, the universe of scientific publications is divided into three main branches and their constituent subfields: science and engineering (with 171 subfields; some major subfields include medicine, biology, and materials science), arts and humanities (with 27 subfields; some major subfields are literature, religion, art, and architecture), and social sciences (which we do not consider in this chapter). The universe of U.S. patents was treated as a separate category (with 36 subfields that include areas such as drugs and medical, mechanical, and electrical patents).[2]

## 3.2 Findings

After analyzing all the data, we found that for science and engineering there has been a substantial increase in collective research/innovation over the years we studied. And, not only are the numbers of collaborative teams increasing but the size

---

[2] A complete list of all sub-categories can be found at
    www.kellogg.northwestern.edu/faculty/jones-ben/htm/Teams.ScienceExpress.pdf. Accessed 2011.

of the teams is increasing as well. Just consider that the Human Genome Project in the year 2000 included a team that consisted of hundreds of scientists at 20 different sequencing centers in China, France, Germany, Great Britain, Japan, and the United States. The Human Genome Project is, in fact, often cited as the break point when life science moved from an individualistic enterprise – with researchers pursuing medical investigations more or less independently – to a collaborative, team-driven discipline. The magnitude of both the technological challenge and the necessary financial investment prompted the Human Genome Project to assemble interdisciplinary teams (encompassing engineering and informatics as well as biology), automate procedures wherever possible, and concentrate research in major centers to maximize economies of scale.[3] Today life science research is characterized by large-scale, cooperative efforts involving many institutions – often from many different nations – working collaboratively.

The trend towards teamwork is now widespread in all areas of science, with team size growing steadily for each of the years we examined from 1955–2000, and authors per paper nearly doubling from 1.9 to 3.5. In addition, the number of research collaborations increased and productivity of these collaborations also improved, as reflected in the data for patents that we examined for the years between 1975 and 2000. That data also shows a definite upward trend in team size during that period when average team size rose from 1.7 to 2.3 inventors per patent. The only area of research where single authors still produce over 90 percent of the papers published is in the arts/humanities.

An overview showing the shift towards collaboration is captured in Table 3.1. In science and engineering, 99.4 percent of the 171 subfields we studied show increased teamwork. At the same time, 88.9 percent of the 27 subfields in the humanities, and 100 percent of the 36 subfields in patenting show an increase in teamwork.

Areas such as medicine, biology, and physics have seen at least a doubling in mean team size over the 45-year period we examined. Surprisingly, even mathematics, long thought to be the domain of the loner scientist and least dependent of the hard sciences on lab scale and capital-intensive equipment, showed a marked increase in the amount of work done in teams, from 19 percent to 57 percent, with mean team size rising from 1.22 to 1.84 persons.

### 3.2.1 Teams and citations

We used the number of citations each paper and patent received, as this has been shown to correlate with research quality (Trajtenberg, 1990; Hall *et al.*, 2005; Aksnes, 2006) and is frequently used in promotion and funding reviews (Ali *et al.*,

---

[3] The Human Genome Project Completion – National Human Genome Research Institute. National Institutes of Health. www.genome.gov/11006943. Accessed 2011.

Table 3.1. *Patterns by subfield. For the two broad ISI categories cited here and for patents, we counted the number (N) and percentage (%) of subfields that show: (1) larger team sizes in the last five years compared to the first five years and (2) RTI (Relative Team Impact) measures larger than one in the last five years. RTI greater than one indicates that teams produce more highly cited research than individuals. We show RTI measures both with and without self-citations in calculating the citations received. Dash entries indicate data not applicable.*

|  | Increasing team size | | | RTI > 1 (with self-citations) | | RTI > 1 (no self-citations) | |
|---|---|---|---|---|---|---|---|
|  | $N_{fields}$ | $N_{fields}$ | % | $N_{fields}$ | % | $N_{fields}$ | % |
| Science and engineering | 171 | 170 | 99.4 | 167 | 97.7 | 159 | 92.4 |
| Arts and humanities | 27 | 24 | 88.9 | 23 | 85.2 | 18 | 66.7 |
| Patents | 36 | 36 | 100.0 | 32 | 88.9 | – | – |

1996). For our purposes, a highly cited work was defined as receiving more than the mean number of citations for a given field and year.[4] Our studies showed that collaborations produced more highly cited work in each broad area of research and innovation at each point in time.

To explore the relationship between teamwork and impact in more detail, we defined the relative team impact (RTI) for a given time period and field. The RTI is the mean number of citations received by team-authored work divided by the mean number of citations received by solo-authored work. An RTI rating greater than 1 indicates that teams produce more highly cited papers than solo authors, and vice versa for an RTI rating less than 1. When RTI is equal to 1, there is no difference in citation rates for team- and solo-authored papers. In our dataset, the average RTI was greater than 1 at all points in time and in all research areas: science/engineering, arts/humanities, and patents.

In other words, the data show that there is a broad tendency for teams to produce more highly cited work than individual authors. And, the relative impact of these collaborations is continuing to rise over time. For example, in science and engineering, team-authored papers received 1.7 times as many citations as solo-authored papers in 1955, but 2.1 times the citations by 2000. Similar upward trends in relative team impact also appear in arts/humanities. The trend, although still upward,

---

[4] Citations received were counted from publication year to 2006. Recent publications have smaller citation counts because they have had less time to be cited, but this effect is standardized when comparing team versus solo publications within a given year.

is a bit weaker in patents where citations can limit the scope of an application and, therefore, are less likely.[5] During the early years that we examined, solo authors received substantially more citations on average than teams in many subfields, especially within science/engineering. By the end of the observation period, however, teams received more citations than individuals in virtually all of the subfields.

The citation advantage of teams has also been increasing with time when teams of fixed size are compared with solo authors. In science and engineering, for example, papers with two authors received 1.30 times more citations than solo authors in the 1950s but 1.74 times more citations in the 1990s. In general, this pattern prevails for comparisons between teams of any fixed size versus solo authors.

A possible challenge to the validity of these observations is the presence of self-citations given that teams have opportunities to self-cite their work more frequently than a single author. To address this, we re-ran the analysis with all self-citations removed from the dataset.[6] We found that removing self-citations can produce modest decreases in the RTI measure in some fields; for example, RTIs fell from 3.10 to 2.87 in medicine and from 2.30 to 2.13 in biology. While removing self-citations can reduce the RTI by five to 10 percent, the relative citation advantage of teams remains essentially intact.

### 3.2.2  Citation impact for solo and team scientists

Because the progress of knowledge may be driven by a small number of key insights (Kuhn, 1970), we further tested whether the most extraordinary concepts, results, and technologies are the province of solitary scientists or collaborative teams. To determine this, we pooled all papers and patents within the areas we researched, then calculated the frequency distribution of citations to solo-authored and team-authored work, comparing the first five years and last five years of our data.

Our results show that teams now dominate the top of the citation distribution in all of our research domains. In the early years, a solo author in science and engineering was more likely than a team to have a work with the highest number of citations, meaning it was a singularly influential paper (the corollary was also true, in that a solo author was more likely than a team to receive no citations). However,

---

[5] In patenting, we may observe weaker trends because (1) citing earlier work can limit a patent's scope, so applicants may avoid citations, and (2) patent examiners typically add the majority of citations, which makes patent citations different from paper citations. See also: Sampat (2005); Alcácer and Gittleman (2006).

[6] A self-citation is defined as any citation where a common name exists in the authorship of both the cited and the citing papers. All citations were removed in which a citing and cited author's first initial and last name matched. This method can also eliminate citations where the authors are different people but share the same name. However, performing Monte Carlo simulations on the data, we find that such errors occur in less than 1 of every 2,000 citations. Thus, any errors introduced by this method appear negligible. We did not remove self-citations from patents because citations to previous work in the patent literature are primarily assigned by the patent examiner, who independently assigns citations to earlier work based on the relevance of previous patents' content. See: Alcácer and Gittleman (2006).

by the last five years of our study, a team-authored paper had a higher probability of being very highly cited. For example, a team-authored paper in science and engineering is currently 6.3 times more likely than a solo-authored paper to receive at least 1,000 citations. When we looked at the areas of arts/humanities and patents, we saw that it was never the case that individuals produced more influential work than teams. These patterns hold true even when self-citations are removed.

It is evident, then, that while solo authors did produce the papers of singular distinction in science and engineering in the 1950s, the mantle of extraordinarily cited work was passed to teams by the year 2000.

### 3.3  Notes on the link between teams and networks

How do you assemble collaborative teams that are destined to be winners? It is a question well worth examining in light of the fact that so many of today's creative ideas, research, and discoveries are occurring in teams rather than in seclusion. To give us some clues, we thought it was important to examine the way teams are being organized in the first place, to see what the most successful groups have in common. One particular industry that comes from outside of the life sciences, but has a lot to teach about managing successful collaborations, is the Broadway musical industry – in part because they have been at it for so long and have created a useful model that determines the number of members and the creative roles needed in the collaboration. A little later in the chapter we will look at the Broadway model to see what implications its success may have for other industries.

Our analysis showed that the make-up of a creative team has a tremendous impact on whether or not it will be successful. And, our analysis also confirmed that the way a team chooses its members determines the kind of larger network the group will become a part of, which in turn plays an important role in how well the team does. Team assembly is vitally important at both the micro-level of individual teams and at the macro-level where these individual teams become part of a large inter connected cluster.

The most successful teams have members who have interconnected webs of colleagues in many fields, in other words, a large network from which to draw ideas. Participation in such a network can offer the kind of inspiration, support, and evaluation among collaborators that promotes the most winning enterprises. But teams can also be part of a number of other types of clusters, many of which are less advantageous to success. Whether in the scientific or artistic areas, the way teams assemble themselves will determine both the structure of their collaborative networks, team performance, and innovative output.

To examine how a successful creative team is assembled, we decided to look at some of the elements that go into the make-up of a team. We identified three areas for consideration: team size, the proportion of newcomers joining a new production, and the tendency for incumbent team members to repeat collaborations with former team members. The latter two categories determine the experience and diversity levels present on a team.

### 3.4 Author networks and team assembly

Teams are created because of the need to incorporate individuals with different ideas, skills, and resources. A team's creativity is enhanced by the varied attributes introduced by its disparate members. This is beneficial in a larger context because fresh thinking in one domain can encourage proven innovations to be introduced into another domain, thereby solving old problems.

But assembling a successful team depends on choosing members who will provide the team with the right balance of diversity and cohesion, which is not quite as easy as it sounds. Research shows that the right balance of diversity on a team is elusive. And, although diversity may potentially spur creativity, it typically promotes conflict and miscommunication (Larson *et al.*, 1996; Edmondson, 1999; Jehn *et al.*, 1999). In general, people feel more comfortable working with others they already know, and choosing new and diverse teammates runs counter to the sense of security most individuals experience when working and sharing ideas with past collaborators (Stasser *et al.*, 1995). The problem, though, is that homogenous teams rarely perform as well as their more diverse counterparts, even with their innate difficulties.

Another important factor to consider is team size. Successful teams evolve towards a size that is large enough to enable specialization and effective division of labor among teammates, but small enough to avoid the overwhelming costs associated with coordinating a large group (Katzenback and Smith, 1993). This evolution to optimum team size is quite evident in our first example, the Broadway musical industry.

In order to determine how winning teams come about, we investigated the mechanisms by which teams of creative agents are assembled. We investigated how microscopic team assembly mechanisms determine both the macroscopic structure of a creative field and the success of certain teams in using the resources available to them in the field. In other words, we started on a very small scale and looked at the way individual teams assemble their membership and then extended our investigation to see how this impacted the overall creative field in which the team was working.

To examine the effects of team composition, we developed a model in which the selection of the team members was controlled by three parameters: (1) the number of team members; (2) the probability of selecting incumbents – that is, individuals already belonging to the network; and (3) the propensity of incumbents to select past collaborators as teammates. Varying these three parameters gave us a wide breadth of possibilities for the way teams can be built, ranging from a small team of all newcomers who have never worked together to a large team composed completely of seasoned, past collaborators; plus many variations in between.

Depending on the values of the three parameters, our model predicted the existence of two different network phases. In one phase, there was a large cluster connecting a substantial portion of the teams; while in the other, the teams formed a large number of isolated clusters. The model confirmed that different kinds of networks will develop depending on the way individual teams combine the three key elements of size, diversity, and experience.

We chose fields to study that experience collaborative pressures such as: differentiation and specialization, internationalization, and commercialization (Ziman, 1994; Brown, 2000; Uzzi and Spiro, 2005). The fields we looked at were: the Broadway musical industry, economics, biology, and astronomy.

For each of the scientific disciplines, we considered all collaborations that resulted in publications in recognized journals within the fields studied: nine economics journals, ten biology journals, and six astronomy journals.[7] Collaboration networks (Newman, 2001; Barabási *et al.*, 2002; Newman, 2004; Borner *et al.*, 2004; Ramasco *et al.*, 2004) were then built for each of the journals independently, then for the whole discipline by merging the data from the journals within a discipline. In the area of Broadway musicals, we analyzed old playbills going back over 100 years.

We wanted to see what we could learn about collaborative teams working to produce musicals, so we considered all 2,258 Broadway productions in the period from 1877 to 1990 (Green and Green, 1996; Simas, 1988). Productions defined as musical shows were those performed at least once on Broadway. Team composition was defined as individuals responsible for composing the music, writing the libretto and the lyrics, designing the choreography, directing, and producing the show. Actors were not included in the calculations.

The evolution of team sizes in Broadway musicals bears out our expectation that team size and composition depend on the intricacy of the creative task. In the period from 1877 to 1929, when the form of the Broadway musical show was still being worked out through trial and error (Green and Green, 1996), there was a steady increase in the number of artists per production, from an average of two to

---

[7] We imposed several requirements on the journals we selected for analysis. Firstly, the main subject category of the journal must be the desired one. For example, we consider only those biology journals whose subject category is either biology or biology and biodiversity, and conservation according to the Journal Citation Reports. We disregarded more specialized journals, such as *Microbial Biology*, whose subject category is more specific. We also required that journals contain a sufficiently large number of papers, typically larger than 1,000.

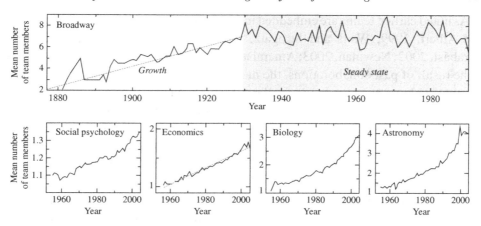

Figure 3.1. Time evolution of the typical number of team members in the Broadway music industry and scientific collaborations in the disciplines of sociology, economics, biology, and astronomy (Guimerà *et al.*, 2005).

an average of seven (Figure 3.1). This increase in size suggests that teams evolved to manage the complexity of the new artistic form. By the late 1920s, the Broadway musical reached the form we know today, as did team composition. The typical set of artists creating a Broadway musical since the 1920s has been choreographer, composer, director, librettist, lyricist, and producer; and for the following 55 years, the average size of teams remained around seven.[8]

### 3.4.1 Team size in the sciences

There are some similarities between Broadway and scientific collaborations with regard to the evolution of team size. To test this, we compared three fields: biology, astronomy, and economics to determine increase in team size over time (Figure 3.1). While the increase has been roughly linear in economics, it has been faster than linear in biology and astronomy. This suggests two things: team size depends on the intricacy of the enterprise, and successful teams may adapt faster to external pressures (such as those mentioned earlier: differentiation and specialization, internationalization, and commercialization).

### 3.4.2 The network

When we looked closely at team size and its evolution, we found that size alone was not a helpful criterion for explaining how teams become part of a larger network.

---

[8] This stationary state remained until the mid 1980s, when size dropped again precisely at the time when a rash of revivals and revues conceivably simplified production.

As noted earlier, teams are embedded in a larger more complex network (Barabási and Albert, 1999; Watts and Strogatz, 1998; Amaral *et al.*, 2000; Albert and Barabási, 2002; Newman, 2003; Amaral and Ottino, 2004). The network that forms is the result of past collaborations, the medium in which future collaborations will develop, and acts as a storehouse for the pool of knowledge created within a particular field. The range of network variations is wide, with isolated clusters operating at one end and large, complex, connected clusters at the other. It is important to note that the type of network in which a team is embedded affects the manner in which the team accesses the knowledge in its field, which has implications for its ultimate level of success.

Teams composed of members with links to many different sets of collaborators are able to draw from a wide and diverse reservoir of knowledge. These individual connections enable teams to connect across a giant cluster of scientists, or artists. The more links an individual has, the more connections are made for the whole team. At the same time, the way teams are organized into a larger network affects the likelihood of breakthroughs occurring in a given field.

The people who make up a team can be classified according to their level of experience. Some individuals are newcomers, that is, rookies, with little experience and unseasoned skills. Others are incumbents; these are established persons with track records, reputations, and identifiable talents. The differentiation of team members into newcomers and incumbents results in four possible types of link within a team: newcomer–newcomer; newcomer–incumbent; incumbent–incumbent; repeat incumbent–incumbent. And the distribution of these different types of link reflects the team's underlying diversity. For example, if a team has a preponderance of repeat incumbent–incumbent links, it is less likely to have innovative ideas because the members' shared experiences tend to homogenize their pool of knowledge. In contrast, a team with a variety of types of links is likely to have more diverse perspectives to draw from, which in turn may contribute to more innovative solutions.

But measuring the emergence and effects of team diversity (Etzkowitz *et al.*, 1994; Harrison *et al.*, 1998; Barsade *et al.*, 2000; Reagans and Zuckerman, 2001; Larson *et al.*, 1996; Edmondson, 1999; Jehn *et al.*, 1999) is more difficult than measuring team size, so we created a model for the assembly of teams. Our model assembled teams in sequence, with each team controlled by three parameters: number of team members, probability of selecting incumbents, and the propensity of incumbents to select past collaborators to work with on new projects.

### 3.4.3 A plausible model

In our model, the number of team members was kept constant for theoretical analysis. Our model's second parameter – incumbency – indicates the level of opportunity for newcomers; the more incumbents on a team the fewer openings for newcomers to enter the field. The third parameter represents the inclination for incumbents to collaborate with prior collaborators, rather than initiate a new collaboration with an incumbent they had not worked with in the past.

Since we wanted to model the emergence of a collaborative network, we built our model starting at time zero with an endless pool of newcomers (newcomers became incumbents after having been chosen for a team the first time). We continued building teams in sequence over time with varying probabilities of selecting members from pools of incumbents, newcomers, and past collaborators.

We then used our model to see if we could predict what kind of network would emerge in a given field. Specifically, we wanted to see if there was evidence of a large connected cluster that comprised most of the agents, or if the network was composed of numerous small clusters. A large connected cluster would support the idea of the invisible college, or web of social and professional contacts that links scientists across universities, as proposed by de Solla Price (1963) and Merton (1973b). Whereas large numbers of small clusters indicate that a field is made up of isolated schools of thought. Our model, in each of the areas we examined, showed a large connected cluster.

There is a point at which the system undergoes what is called a phase transition, that is, it experiences a sharp transition from a multitude of small clusters to a more ordered situation in which one large cluster, comprised of a substantial fraction of the individuals, emerges. The large cluster that emerges – the so-called giant component – then begins acting as a single entity. When this transition will occur in a network correlates with the probability of selecting incumbents (people already operating within the network) as team members. The more incumbents included at each time stage when new teams are being formed means more and more links are created until the largest cluster eventually forms a densely connected network.

For each of the fields for which we had empirical data, we measured the relative size of the giant component. For all of the fields, the giant component was greater than 50 percent – the point at which an invisible college emerges. Our model, which correctly predicted these results, provides quantitative evidence for the existence of an invisible college (linking scientists across universities) in all of the fields we studied.

### 3.4.4 Assembly rules and journal impact factors

If diversity affects team performance, and our model correctly captures how diversity is related to the way teams are assembled, then the qualities of diversity and incumbency/expertise must be related to team performance. In order to investigate this in the scientific fields, we considered how real-life teams that publish in different quality journals are assembled. We used each journal's impact factor as a proxy for the typical quality of a team's output. We then studied the different journals separately to quantify the relationship between how teams are created and their performance and innovative output.

We found that in the areas of economics, biology, and social psychology, incumbency was positively correlated with a journal's impact factor. This implies that successful teams have many incumbent members who provide the group with expertise and know-how. Diversity was negatively correlated with impact factor; meaning that less diverse teams, those composed of incumbents who mainly select to work with prior collaborators, have lower levels of performance. While these kinds of team alignments may breed familiarity, they do not breed the ingenuity that adding new team members can spur.

Teams that publish in journals with a high impact factor, that is, well-respected journals, typically give rise to a large giant component. Teams that publish in highly respected journals are ones that perform a better sampling of the knowledge within their field and are able to more efficiently use the wide array of resources of the invisible college for their benefit. Teams that form small isolated clusters tend to produce work that is of lesser significance; hence, publishing in less well-regarded journals.

Surprisingly, neither expertise, diversity, nor the size of the network's giant component could be significantly correlated with impact factor in the field of astronomy.

Our analysis leads us to conclude that discoveries, innovations, and new ideas belong more and more to the realm of teams rather than the single practitioners of old. Today's high quality research is being done in teams across nearly all fields, from the arts to the sciences. Following this, we state what we learned about the three team variables we studied: size, expertise, and diversity.

We found that team size is evolving with time, probably up to an optimal size – as in the case of the Broadway music industry – and it is our contention that a similar process may also be occurring with regard to expertise and diversity.

We found that a strong correlation exists between the way teams choose their membership and the quality of the work they produce. While experience is important, successful teams choose enough new people who can provide a creative spark and stimulate new ideas and approaches. Too many incumbents, especially those who have worked together repeatedly in the past, limits diversity, which can lead to low impact research or productions. Unsuccessful teams repeat past collaborations over and over, thereby denying themselves fresh thinking. Successful teams create a balance of new and old participants by bringing in experienced people who have not worked together before.

We also discovered that in the sciences, successful collaborations tend to produce innovations and high quality research that is published in the most respected journals. In addition, successful teams are embedded into a large social network that provides them with many contacts across fields, and the more links they have, the more extra links they are likely to beget.

The fact that in one of the fields we observed, astronomy, there are no correlations between expertise, diversity, or cluster size, and the quality of journal publication indicates that this field is different from the others. Whether these differences are caused by the needs imposed by the creative enterprise itself, or to historical or other reasons is a question that we cannot answer conclusively at this time.

However, for the remaining fields we studied – Broadway musicals, economics, and biology – very similar values of expertise and diversity are evident. This suggests that a universal set of optimal values might exist; values that may lead us to understand just the right mix for assembling successful teams whether they are on the stage or in the laboratory.

## 3.5 Discussion

The innovative process in the sciences has undergone a fundamental shift, as evidenced by the universally increasing number, size, and complexity of teams over the last 15 years. This highly intertwined team structure has unleashed enormous innovation and creativity in the life sciences. The demands for interdisciplinary knowledge and expensive equipment make the need for teamwork critical to success. And an efficient team structure becomes even more essential in helping to maintain competiveness in the marketplace, as the life science sector persists in growing more complex and interdisciplinary.

Certainly, under these circumstances, the need to assemble highly efficient and impactful teams will continue to grow and increase in significance. Even so, the idea of *team* is a relatively new concept in the life science sector and there is a lot to be gained by looking at the lessons learned from other, older disciplines such as Broadway, humanities, and the basic sciences that point to the importance of size, expertise, and diversity as central considerations in assembling a winning team.

# 4

# Structural folds: the innovative potential of overlapping groups[1]

BALÁZS VEDRES AND DAVID STARK

Entrepreneurial groups face a twin challenge: recognizing new ideas and implementing them. Recent research suggests that connectivity reaching outside the group channels new ideas, while closure makes it possible to act on them. By contrast, we argue that entrepreneurship is not about importing ideas but about generating new knowledge by recombining resources. In contrast to the brokerage-plus-closure perspective, we develop a concept of structural folding and identify a distinctive network position, structural fold, at the overlap of cohesive group structures. Actors at the structural fold are multiple insiders, participating in dense cohesive ties that provide close familiarity with the operations of both groups. Structural folding provides familiar access to diverse resources. Firstly, we test whether structural folding contributes to higher group performance. Secondly, because entrepreneurship is a process of generative disruption, we test structural folding's contribution to group instability. Thirdly, we move from dynamic methods to historical network analysis and demonstrate that coherence is a property of interwoven lineages of cohesion that are built up through an ongoing pattern of separation and reunification. Business groups use this pattern of interweaving to manage instability while benefiting from structural folding. To study the evolution of business groups, we construct a dataset that records personnel ties among the largest 1,696 Hungarian enterprises from 1987–2001.

[1] This chapter is based on an earlier publication in the *American Journal of Sociology* (Vedres and Stark, 2010). Research for this chapter was supported by a grant from the National Science Foundation #0616802. We are alternating the order of authors' names across a series of papers produced from our broader project on historical network analysis. For their comments, criticisms, and suggestions we are grateful to Peter Bearman, Laszlo Bruszt, Monique Girard, David Lazer, David I. Levine, Anna Mitschele, Guido Moellering, John Padgett, Roger Schoenman, and Tamás Vicsek. Our thanks to the Institute of Advanced Study, University of Durham, UK, and to the Max Planck Institute for the Study of Societies in Cologne for hosting Stark as a Visiting Fellow while the manuscript was in preparation.

Business groups face two key challenges in their entrepreneurial mode of operation: to recognize sources of novel ideas and to secure the means to implement them. Recent thinking suggests that brokerage ties of connectivity outside the group provide contact with new ideas in the environment; and that cohesive ties of closure within the group provide trust and mutual understanding for implementation (Burt, 2005; Uzzi and Spiro, 2005; Obstfeld, 2005). Whereas this brokerage-plus-closure perspective sees innovation as importing and implementing ideas, we offer an alternative conception of entrepreneurship, as recombination. In our view, truly innovative ideas – in the first instance, a fresh conceptualization of the problem itself – are not free-floating outside the group.

Instead of importing ideas or information, the challenge is to generate knowledge. It follows that the work of recombination that generates new knowledge requires intense interaction and deeply familiar access to knowledge bases and productive resources, as opposed to long-distance contact and casual access. From this perspective, entrepreneurship, as an enabling capacity, proves productive not so much by encouraging the smooth flow of information or the confirmation of fixed identities, but by fostering the generative and productive friction that disrupts the received categories of "business as usual" and enables the redefinition, redeployment, and recombination of resources (Stark, 2009). Yet, a simple expansion of cohesive group membership would not be sufficient: recombination requires interaction across diversity.

Therefore, we argue that entrepreneurship in the business-group context is driven by the intersection of cohesive groups where actors have *familiar* access to *diverse* resources available for recombination. In making this argument, we draw on Simmel's ([1922] 1964) insight that membership in cohesive groups can overlap. With a method that allows us to identify cohesive – yet nonexclusive – groups, we bring theoretical attention to the distinctive structural position at their intersection. Our concept of *structural fold* is a theoretical counterpart to Burt's (1992) concept of "structural hole" with different network properties. Corresponding to our different understanding of the innovative process and the structural basis for it, structural folding, as mutual participation in multiple cohesive groups, provides the requisite familiarity and diversity for access and for action through a distinctive network topography that is not a summation of brokerage and closure.

We further argue that entrepreneurship has not only structural properties within a synchronic dimension, but also dynamic properties along the temporal dimension. Specifically, entrepreneurial structures are not only creative but are also likely to be disruptive. Thus, our dual task is to analyze structural features that predict group performance *and* to analyze whether and how these same structural features contribute to or undermine the continued existence of the group itself.

Postsocialist Hungary offers an excellent case for examining organizational innovation and the historical evolution of business groups. The dislocations and uncertainties facing state-owned enterprises undergoing privatization, new start-ups, and foreign-owned subsidiaries can hardly be overstated. With the collapse of COMECON (the trading system among Soviet-bloc states) following the demise of the Soviet Union, many firms faced the almost total collapse of their once-guaranteed markets. Business networks and business groups emerged quickly as organizational experimentation rapidly gathered speed to locate or generate markets for products of the group and to buffer members from varied uncertainties. Several studies have demonstrated the innovative capacity of postsocialist firms and groups in making use of new corporate institutions, such as cross-ownership, board ties, and holding structures (Stark, 1996; Spicer *et al.*, 2000).

Our dataset documents this process of organizational experimentation and the dynamics of group formation and entrepreneurship. We have collected the names of all economic officeholders in Hungary from 1987 to 2001, defining economic officeholders as all senior managers and members of the boards of directors and supervisory boards of the largest 1,696 companies. With our list of 72,766 names and the exact dates of their tenure on these boards, we can construct the personnel ties connecting these largest firms for each year in our study. Our case reaches back to the very moment when firms could adopt the newly legalized, corporate form. It includes periods of business uncertainty involving privatization, transformational recession, marketization, the institutionalization of economic regulations, massive foreign direct investment (FDI), as well as three parliamentary elections.

## 4.1 Network structures for access and action

One prominent feature of business networks is that firms cluster together in cohesive groups. Dense ties among the members of the group, it is argued, provide a basis for trust and a means for coordinating action (Useem, 1980; Uzzi, 1997). Cohesive ties enable groups to implement projects beyond the capacity of any given firm (Granovetter, 2005). The sharing of risk along these ties buffers groups from uncertainty. Because it mitigates the impact of abrupt downturns, risk-spreading in such business-group networks lowers the volatility of year-to-year profitability (Lincoln *et al.*, 1996).

Another logic states that business groups might elect to forego high density within the group in favor of maintaining more weak ties to firms outside the group. Such a strategy of sacrificing density for diversity economizes network resources by reducing the number of redundant ties (Burt, 1992). Long-distance ties provide access to more channels of information outside the group, and this diversity provides a basis for greater adaptability. According to this logic, the conserving strategy of in-group cohesion is maladaptive. This strategy runs the risk of locking the business group into early successes, a strategy that will quickly become detrimental in a situation of rapid change in which the directionality of disruption cannot be foreseen. The defensive strategy of closing ranks risks the chance of creating a false sense of security at a point when the actual situation does not call for pitting survival and innovation against one another, but for seeking innovation in order to survive.

Recent developments in network analysis suggest a third strategy: in place of strengthening ties within the group *or* reaching outside it, do both. Some researchers use the terms cohesion *and* connectivity to characterize this strategy (Watts, 1999; Moody and Douglas, 2003; Uzzi and Spiro, 2005); others favor the terminology of closure *and* brokerage (Burt, 2005, 2008; Baum *et al.*, 2007). Common to all is the notion of the complementarity of these distinctive network properties, regarded as especially beneficial in cases where the goal is innovation. Brokerage/connectivity provides access to ideas and information but in itself lacks the means for implementation. Closure/cohesion provides the means of coordination but lacks diversity for discovery. Together, they can compensate for the limitations of each; Obstfeld (2005) has labeled these as the "action problem" and the "idea problem." Exemplary of this approach is the recent study by Uzzi and Spiro (2005), who demonstrate that the success of Broadway musicals (in which innovation is the ability to produce a "hit") is a function of enough cohesion (the continuity in the composition of the musical "team" from one musical to the next) and connectivity (diversity in composition from one musical to the next).

Our study of business groups in Hungary takes these recent developments as its point of departure. With Burt, Uzzi, and others, we share the notion that entrepreneurship (or, more generally, innovation) is not facilitated by either closure/cohesion or brokerage/connectivity alone. Because we agree that the innovation involves a combination of close familiarity and diversity, we similarly seek to identify specific structural features that promote these social processes. However, we depart from the current consensus by arguing for a perspective on entrepreneurial structures that does not involve a summation of the conventional dichotomies. Instead, as we elaborate theoretically in the following section, we develop a conception of *structural fold*, a distinctive network structure built from intersecting cohesive groups.

Because we see entrepreneurship as disruptive as well as creative, we are attentive to the finding of Uzzi and Spiro (2005) that innovation requires some reshuffling of groups. Very stable groups (a Broadway musical formed with the same members who worked together in a previous musical) and highly unstable groups (a musical in which few members had worked together before) are less likely to produce an innovation than groups combining members with prior affiliation and novel affiliations. With this insight in mind, we take the next analytical step: is there a structural feature that can predict successful performance and also explain the dynamics of group formation and dissolution?

## 4.2 Structural folds

To address this question we must first identify the relevant groups in network terms. Our case differs from those of Burt and Uzzi, for whom group boundaries (e.g. the members of a team project, the members of a musical) are given prior to the analysis. Business groups in Hungary – unlike those in East Asian economies, where analysts can refer to directories listing the members of Japanese *keiretsu* or South Korean *chaebol* – do not exist as named places on the economic landscape.

We adopt a measure of cohesion to identify the components of business groups, using the guiding theoretical principle that cohesive group structures need not necessarily be exclusive. That is, we are deliberately attuned to the possibilities that network structures can be cohesive *and* overlap. As Simmel ([1922] 1964) observed in the *Web of Group Affiliations*, a person is frequently a member of more than one cohesive group at a given time. For Simmel, such multiple-group membership was a source both of individuation for the person and of social integration for the larger collectivities involved. In our population of firms in Hungary we might find, for example, a power plant that is linked to other power plants in a cohesive group, while also cohesively tied with power distributors and coal mines, linked to other heavy industry companies in a different cohesive group, and associated in a group with banks.

We develop a concept, structural folding, to refer to mutually interpenetrating, cohesive structures. Figure 4.1 illustrates the structure of structural folding in contrast to brokerage and closure. It also shows the distinctive network position, the structural fold at the intersection.[2] Actors at the structural fold are multiple

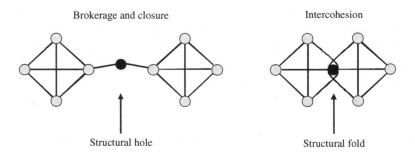

Figure 4.1. The structure of structural folding in contrast with brokerage and closure.

---

[2] Recent work in social network methodology acknowledges that actual network structures can be composed of overlapping cohesive groups (Moody and Douglas, 2003). Our analytical contribution is to recognize that, if cohesive groups can overlap, there is a distinctive structural position at the intersection. That is, from a methodological residual, we point to the intersecting location as a sociological object worthy of theoretical reflection.

insiders, participating in dense cohesive ties that provide close familiarity with the operations of the members in their group. Because they are members of more than one cohesive group, they have familiar access to diverse resources. This combination of familiarity and diversity facilitates the work of recombining resources. As the node that is common to multiple groups, structural folds are resources for the groups themselves. Structural folding is closure without being closed off; it is generative cohesion without insularity.

Our identification of distinctively intercohesive processes (as opposed to an additive operation of the structure of internal group ties and the structure of external group ties), rests on conceptual differences about entrepreneurship. For the adherents of the additive school, innovation is basically conceptualized as a process of germination: bridging or brokerage ties bring the seeds of ideas and information to the nurturing soil of trusting relationships of cohesion. We have no doubts that ideas and information are often vital; our question is whether the activity of productive recombination involves only (or even primarily) resources that can flow or circulate.

With Schumpeter, we conceptualize entrepreneurship as recombination,[3] a process more complex than the importation and implementation of novel ideas. In our view, the *idea problem* is itself an *action problem*. That is, the most innovative ideas are not "out there" in the environment of the group. Instead of waiting to be found, they must be generated (Kogut and Zander, 1992). It is one thing to recognize an already-identified pattern, but quite another to make a new association. In this sense, the process of innovation is paradoxical, for it involves a curious cognitive function of recognizing what is not yet formulated as a category (Stark, 2009). As John Dewey ([1938] 1998) and the pragmatists argued, it is only in the process of attempting to make a transformation in the world that new problems can even be formulated. Generating novel recombinations is itself a kind of production requiring coordination and cooperation across different communities.

In their study of new product development in cellular telephones, blue jeans, and medical devices, Lester and Piore (2004) demonstrate that each of their cases of radical innovation involves combinations across disparate fields: fashion jeans are the marriage of traditional workmen's clothing and laundry technology borrowed from hospitals and hotels; medical devices draw on both basic life sciences and clinical practice; and cellular phones recombine in novel form radio and telephone technologies. They conclude that "without integration across the borders separating

---

[3] We are drawing on Schumpeter's definition of entrepreneurship as "the carrying out of new combinations." "As a rule, the new combinations must draw the necessary means of production from some old combinations ... development consists primarily in employing existing resources in a different way, in doing new things with them" (Schumpeter, 1934, p. 68). For a neo-Schumpeterian statement in the field of economic growth models, see Weitzman (1998).

these different fields, there would have been no new products at all" (Lester and Piore, 2004, pp. 14–15).

For us, the telling phrase in this passage is "integration across the borders..." Lester and Piore do not refer to "contacts" across borders, for it is not enough for different communities to be *in contact*. Recombinant innovation requires that they *interact*. For these reasons, the conventional pairing of access to diversity is insufficient. Deep access for generating new problems, new knowledge, and new capabilities (as opposed to transferring already accepted ideas) requires considerable trust, hence familiarity. Such access can only be achieved through insiders, accepted members of groups. Therefore, we argue that productive recombination at the group level requires familiar access to resources that are not provided by the narrow bandwidth of the slender ties of bridging and brokerage.[4]

Yet these ties cannot be so cohesive as to constitute a single group: to be able to recognize the potential for novel recombinations, entrepreneurship requires access to *diverse* sets of resources – access that is only possible by being a member of two or more cohesive groups. The distinctive properties of the structural fold provide mechanisms for achieving the diverse familiarities required for recognizing resources and for their productive recombination.

To test these ideas, we correlate the performance of groups and the extent to which they are intercohesive. We expect that groups with more structural folds will perform better than more insular ones.

### 4.3 Instability and coherence

Establishing correlations between structures and performance in a cross-sectional context leaves unaddressed the question of durability: are the ties transitory or are they indicative of a sustainable, self-generating pattern?[5] While testing whether structural folding is performance enhancing, we further analyze whether this same structure is self-perpetuating or self-destructive. From a theoretical perspective, structural folds are points of tension where multiple routines of operation and schemas to organize resources are at work. As prominent locations of restructuring

---

[4] Bridging can only have as great an impact on cohesive groups as a two-step path length. Members of bridged groups are, at best, friends of friends. The strength of impact between groups is limited by the weak ties that are between them. Whereas actors at the structural hole occupy a brokerage position at the gap and tax flows, intercohesive actors occupy an entrepreneurial position at the overlap and recombine resources.

[5] Founding statements in the field of network analysis were attentive to issues of duration and temporality. Moreno and Jennings (1937, p. 371), for example, defined cohesion as "the forces holding the individual within the groupings in which they are." At mid-century, Festinger and Schachter's (1950, p. 164) study of social pressures in informal groups addressed "the total field of forces which act on members to remain in the group" (see McPherson and Smith-Lovin, 2002 and Friedkin, 2004 for discussion). In this emphasis they echoed Simmel, whose publication in an early issue of the *American Journal of Sociology* was entitled, "The Persistence of Social Groups" (Simmel, 1898).

agency, such intersecting social structures can be engines of social change from within (Sewell, 1992).

Our examination of the role of structural folding in the dynamics of group evolution stems from our conception of entrepreneurship. As observed by Schumpeter (1934), entrepreneurship, while fostering innovative recombinations, also contributes to "creative destruction." Expressed in network analytic terms, entrepreneurial structures are likely to destabilize groups. As Uzzi and Spiro's (2005) findings suggest, stability in itself is not the most favorable outcome; indeed, some disruption can be beneficial. We therefore test not only whether the creative tensions of intercohesive familiarity and diversity are performance enhancing in the first instance, but also whether these same tensions foster a creative disruption that disperses group members, who become available for later regrouping. We expect that recombinant opportunities provided by structural folding will have performance-enhancing effects at the group level. We further expect that these benefits come at the cost of group stability.

Overlapping membership can be disruptive for group coordination, relations of reciprocal trust, and a sense of fairness. Those with multiply cohesive attachments might seem to follow strategies that are not transparent to those with single membership, a situation that may hinder coordination. In the face of the ambiguous loyalties of the structural fold, other members might suspect they are being exploited or manipulated; the lack of commitment and time devoted by those with multiple memberships to a given group might lead to group-level dysfunction, perhaps even fracture and fragmentation. Structural folding, we therefore expect, will be negatively correlated with cohesive group stability.

## 4.4 Data and methods
### 4.4.1 Data

The dataset that we have assembled includes the complete histories of personnel ties among the largest enterprises in Hungary spanning the years 1987–2001. We define large companies as those listed in the annual ranking of the top 500 firms (based on revenue) for any year from 1987 to 2001. Our inclusion rule results in a population of 1,696 firms. This population of firms represents more than a third of employment, about half of Hungarian GDP, and almost all export revenues (Figyelő, 2002).

We define *economic officeholders* as senior managers and members of the boards of directors and the supervisory boards of these large enterprises. Personnel data on economic officeholders were transcribed directly from the official files of the twenty Courts of Registry where Hungarian firms are obliged to register information about ownership and personnel. Our dataset on economic officeholders contains 72,766 names.

Beyond economic officeholders, we have also collected the names of all *political officeholders* in Hungary during this same period. For the years 1990–2001, we define political officeholders as every elected politician from the Prime Minister, to the Members of Parliament, to the mayors of all municipalities, including the top three levels in the hierarchy of the national government ministries. For the period prior to free elections, we define political officeholders as all members of the Politburo and the Central Committee of the Hungarian Socialist Workers Party, as well as government ministers and their deputies.[6] Our dataset on political officeholders covering the period from 1987–2001 contains 16,919 names.

We define two companies as having a *personnel tie* if a manager or a board member of one company sits on the board of another company. We record personnel ties as symmetrical, with a starting and ending date for each tie. The models in this study use an annual time resolution, with personnel ties recorded at the last day of each year. We define a firm as having a *political affiliation* when one of its economic officeholders is also a current or former political officeholder,[7] and we record the party affiliation of that political officeholder as the party affiliation of the firm. Political affiliations are thus personnel ties connecting firms and parties.[8]

For each firm, we also collected data on its annual revenues, capitalization, employment, industrial classification, privatization history, and types of owners (state, domestic private entity, or foreign owner).

To identify cohesive groups we adopt a method that starts from cohesive localities, recognizes groups independent of the global network environment, and identifies structural folds. We use the clique percolation method (CPM) developed by physicists to uncover the overlapping community structure of complex networks (Gergely *et al.*, 2005), a method recently demonstrated as a suitable tool to analyze the evolution of cohesive groups (Palla *et al.*, 2007).

---

[6] Data on political and government officeholders were collected from the National Bureau of Elections (which holds records on all elected political officeholders) and from the Hungarian News Agency (which maintains records on all government officials entering or exiting office). Whereas the Communist Party's Central Committee is analogous to the parliament of the subsequent democratic period, the Politburo was akin to the role of the government in the later period. Names of political officeholders in these years from 1987–89 were gathered from a comprehensive CD-ROM publication (Nyírő and Szakadát, 1993).

[7] The motivation to include former as well as current political officeholders comes from our interviews with managers of large firms. As one CEO noted: "In Hungary there is no such thing as an ex-politician. Once a politician, always a politician."

[8] We do not record a personnel tie between two firms when the tie is created by a political officeholder. Two firms might invite the same politician or ministry official to sit on their boards, not because they wish to establish a tie to one another, but because they seek a political and/or government connection. Political affiliations are about personnel ties between parties and firms and not between firms. Including a personnel tie between these firms would introduce noise in the data that would potentially blur the patterns of personnel ties created to foster business collaboration.

## 4.5 Group performance

If recombination though structural folding contributes to entrepreneurship and innovation at the group level, groups with more structural folds should outperform exclusive ones. Does structural folding contribute to group performance? To answer this question we first need to identify a suitable metric of performance and then isolate the contribution of structural folding to performance.

Profitability, although a widely used indicator of performance, has questionable validity in the postsocialist setting. With high taxes and changing government regulations, profits can easily be manipulated – or, euphemistically speaking, "optimized" – depending on prevailing regulations. At a time when most firms were undergoing restructuring programs, profitability was not a practical metric to gauge firm performance. Instead of turning immediate profits, the key to survival was obtaining and keeping markets and thus securing revenues. We choose therefore to focus on revenue dynamics.

We consider the simplest test to be the correlation between structural folding at year $t$ and revenue growth from year $t$ to $t + 1$. For a given group in a given year, structural folding is measured by the number of structural folds, i.e. the number of groups with which that given group overlaps. Revenue growth is the growth rate at group level measured in real terms (corrected for inflation). This correlation between structural folding and revenue growth is significant (with $p = 0.049$) and positive, although weak ($R = 0.088$). This indicates that groups with more structural folds grew slightly faster.

Does higher revenue growth in groups with structural folding accrue only at the structural fold? If structural folding operates similarly to brokerage, we expect that firms at the intersection will see a faster increase in their revenues, compared with firms with only one membership. To answer this question, alongside computing the correlation between structural folding and revenue growth at the group level, we also computed the correlation between the number of cohesive group memberships and revenue growth at the individual level. This Pearson correlation is not statistically significant ($R = 0.039$, $p = 0.127$), which indicates that the benefits of structural folding are realized at the group level and do not appear distinctly at the level of the individual firms occupying the structural fold. The mechanisms of structural folding seem to differ from gatekeeping and brokerage, where benefits – additional revenue-generating possibilities – would accrue at the gatekeeping firm spanning the two groups.

To further test the relationship between growth and structural folding and to isolate this relationship from other predictors of revenue growth, we construct multivariate models (Table 4.1 provides descriptions of the variables that we use). Our goal is to evaluate whether there is a significant relationship between structural folding and growth even if we introduce variables of intracohesive processes

Table 4.1. *Descriptive statistics.*

| Independent variables | Mean | Sd | Min | Max |
|---|---|---|---|---|
| Structural folding | 2.707 | 2.746 | 0.000 | 19.000 |
| Group stability from $t$ to $t + 1$ | 0.514 | 0.259 | 0.120 | 1.000 |
| Group stability from $t - 1$ to $t$ | 0.438 | 0.282 | 0.000 | 1.000 |
| Revenue growth from $t$ to $t + 1$ | 1.587 | 3.945 | 0.170 | 63.190 |
| Negative growth from $t$ to $t + 1$ | 0.544 | 0.499 | 0.000 | 1.000 |
| Top quartile growth from $t$ to $t + 1$ | 0.249 | 0.428 | 0.000 | 1.000 |
| *Intracohesive processes* | | | | |
| Group size | 4.661 | 1.168 | 4.000 | 11.000 |
| Capital size of the largest firm | 9.355 | 1.241 | 1.000 | 10.000 |
| Size difference | 1.507 | 1.650 | 0.000 | 9.000 |
| Financial members | 0.722 | 1.101 | 0.000 | 8.000 |
| Industry homogeneity | 0.409 | 0.317 | 0.000 | 1.000 |
| *Extracohesive processes* | | | | |
| Brokerage | 19.242 | 10.870 | 0.000 | 44.000 |
| Bridging ties | 11.217 | 11.806 | 0.000 | 72.000 |
| State-owned proportion | 0.255 | 0.267 | 0.000 | 1.000 |
| Foreign-owned proportion | 0.253 | 0.231 | 0.000 | 1.000 |
| Politicized proportion | 0.223 | 0.182 | 0.000 | 0.800 |
| Political mix | 0.260 | 0.439 | 0.000 | 1.000 |
| Governing party tie | 0.910 | 1.007 | 0.000 | 5.000 |
| *Controls* | | | | |
| Year | 96.594 | 2.903 | 89.000 | 101.000 |
| Group age | 2.817 | 1.820 | 1.000 | 11.000 |
| Newly formed group | 0.113 | 0.317 | 0.000 | 1.000 |
| Labor efficiency (log) | 0.700 | 0.530 | −0.550 | 2.824 |
| Capital efficiency (log) | −2.636 | 0.652 | −4.513 | 0.447 |
| Sum of revenues (log) | 4.067 | 0.701 | 2.400 | 5.701 |
| *Industry* | | | | |
| Energy | 0.048 | 0.213 | 0.000 | 1.000 |
| Mining | 0.015 | 0.121 | 0.000 | 1.000 |
| Chemical | 0.151 | 0.466 | 0.000 | 4.000 |
| Metallurgy | 0.061 | 0.253 | 0.000 | 2.000 |
| Heavy industry | 0.400 | 0.836 | 0.000 | 5.000 |
| Light industry | 0.331 | 0.680 | 0.000 | 3.000 |
| Wood and textile | 0.140 | 0.396 | 0.000 | 3.000 |
| Food industry | 0.627 | 1.104 | 0.000 | 8.000 |
| Construction | 0.258 | 0.878 | 0.000 | 6.000 |
| Wholesale | 0.421 | 0.655 | 0.000 | 4.000 |
| Retail | 0.367 | 0.659 | 0.000 | 4.000 |
| Transport | 0.102 | 0.319 | 0.000 | 2.000 |
| Services | 0.650 | 0.869 | 0.000 | 4.000 |

(such as group size or homophily), extracohesive processes (primarily those of brokerage and bridging), and control variables, such as industry composition and efficiency.

For each group in our dataset, we have a score of real revenue growth. From this continuous variable we construct two categorical dependent variables capturing the distinctive processes of revenue decline and growth. Groups that are more entrepreneurial should achieve higher performance, but entrepreneurship does not guarantee performance. Thus we also expect that more entrepreneurial groups are not buffered from declining performance.

*Revenue decline* records whether the revenues of the group declined during the year in question. *High revenue growth* records whether the group belonged in the most successful 25 percentile of groups in the overall sample (revenues for this top quartile corresponded to at least eight percent annual growth, controlling for inflation). We have two reasons for transforming our ratio-scale variable into two categorical variables. The technical reason is that the distribution of revenue growth is highly skewed – there are many groups with modest growth and few with extremely high growth. The second reason is substantive – we believe that the predictors of preventing revenue loss and the predictors of achieving high performance are distinct.

Our first independent variable is structural folding: the number of structural folds. We expect structural folding to lower the probability of revenue decline and to increase the probability of high revenue growth.

Our second set of independent variables represents intracohesive processes. The first variable is *group size*, the number of firms in the group. Similarly to Simmel's sociology of numbers, we expect that larger groups will have different revenue dynamics than will smaller ones, as it is less likely that a large group achieves extraordinary growth. The second variable registers processes of homophily based on homogeneity in the industry profile.[9] We measure this *industry homogeneity* by the numerical difference between the first- and second-most prominent industry categories in the group. If the group is entirely made up of one industry, this variable is equal to one. If there are two equally represented industry categories, this variable is equal to zero. Three other variables refer to processes of economic power and dominance in stabilizing or destabilizing the group. *Size of the largest firm* is measured in deciles of capitalization, ranging from 1 (smallest firms) to 10 (largest

---

[9] Homophily has been shown to be an important factor contributing to cohesive affiliation (McPherson *et al.*, 2001). McPherson and Smith-Lovin (2002, p. 13) define homophily as "the positive relationship between similarity (on almost any dimension) and the probability that two people will have a network connection between them." Although McPherson has not studied business groups, the most pertinent dimension of similarity in this context is homogeneity of industry profile. If homophily is operating among our Hungarian firms, cohesive groups concentrated in the same industry should be more likely to exhibit stability than those of greater industrial heterogeneity.

firms). The expectation is that powerful economic players can hold a group together
(Thye *et al.*, 2002). To assess the effects of relative economic dominance, we record
*size difference* as the size-decile difference between the largest and second largest
firms in the group. A larger value indicates a more clearly dominant player in the
group in terms of size. Whereas Thye *et al.* (2002) expect that the equality of power
fosters group stability, Gould's (2003) formulation would predict that equality of
power would lead to conflict and group breakup. The variable *financial members*
records the number of financial firms that are members of a given group. Following
Mizruchi and Stearns (1988), we expect groups with financial members to have
inferior performance.

Our third set of independent variables represents extracohesive processes. For
each group we record *being brokered* as the number of other groups to which the
group is connected by an intermediary. We also record *bridging ties*, the number of
other groups that are reachable with a direct tie from the group in question. These
variables represent brokerage as recently reformulated by Ronald Burt (Burt, 2005).
Whereas Granovetter (1973) would expect that groups with bridging ties will have
higher performance, we are also attuned to the possibility that being brokered has
negative implications on performance (Fernandez-Mateo, 2007).

Among extracohesive processes, we also consider the reach of the group into the
political field. This dimension is salient because our case involves profound economic dislocation in the context of a simultaneous political transformation (Stark
and Bruszt, 1998). Business groups are seen as especially suitable vehicles for political affiliation in such emerging markets (Khanna and Rivkin, 2001). We include
three variables to tap various aspects of these processes. *Politicized proportion*
records the proportion of group-member firms having party affiliations through
personnel ties. A group exhibits *political mix* when we find affiliations to parties on
both the left and the right. *Governing party ties* records a group having a political
affiliation to a currently governing party.

The final variables in this set of extracohesive processes involve links to owners
outside the group.[10] *State-owned proportion* records the proportion of group members for which the state is a significant owner. *Foreign-owned proportion* similarly
records the proportion of group members in foreign ownership.

As control variables, we include *year* as well as *group age*, defined as the average number of years that pairs of group members have spent in groups together.
We include specific *industry* categories, an indicator of whether the group was
*newly formed* by firms that had not belonged to any groups in the previous year,

---

[10] Our dataset contains detailed information about firms' ownership structure. For each firm we can record whether
it has sizable state ownership and sizable foreign ownership, as well as details in the timing of any changes in
such ownership. Our definitions of significant state and foreign ownership follow procedures detailed in Stark
and Vedres (2006).

*labor efficiency* (measured as revenues over number of employees), and *capital efficiency* (measured as revenues over capitalization). To correct for skewness of the distribution on these latter two variables, we take the logarithms.

Both models – the model of declining performance and the model of high performance – pass the Hosmer–Lemeshow test for fit and Pregibon's link test of model specification. Tolerance and variance-inflation factors were within conventional bounds for all independent variables, indicating that multicollinearity should not be a concern. To test for sensitivity in defining the cut point of high performance at the top 25 percent, we ran the high performance model with dependent variables representing the top 30, 20, 15, 10, and 5 percent of revenue growth. In all of these models, the sign and significance of the structural folding coefficient was the same.

In Table 4.2 we see that structural folding does not buffer against revenue decline, but it is a strong predictor of high growth. It should be kept in mind that, by definition, all the groups we examine in Table 4.2 are cohesive. Intercohesive groups outperform their counterparts, who lack this ambiguous yet recombinative advantage.

Turning to intracohesive variables, we see that larger group size makes decline less likely without contributing to growth. For business groups, there is safety in numbers – smaller groups are more at risk of decline. Having a dominant member of large size makes it more difficult for a group to achieve high growth, probably because, for large firms, a high rate of growth means a large increase in revenue volume. Homophily is a disadvantage. Industry homogeneity increases the probability of decline and decreases the probability of high performance. Groups of more heterogeneous composition are advantaged. This does not apply, however, to groups that include members from finance. Groups with financial members are significantly more likely to face declining revenues. This result might not be surprising to scholars who study business groups: ties to banks are often associated with financial troubles and decreasing performance.

Looking at personnel ties reaching to other business groups, we see that effects on performance diminish as distance increases. Groups with more two-step ties to other groups (mediated through brokers) are slightly less likely to decline but show no advantage in high performance. Groups with ties at the closest reach (not even one step away but established through the interpenetrating ties of structural folding) outperform groups that are exclusive, regardless of how intensively or extensively they are embedded. The number of bridging ties to members of other groups decreases the probability of high performance. This finding stands out in sharp contrast to our findings on structural folding. Bridging ties – circuits for the circulation of ideas – do not contribute to group success. This starkly contrasts the positive contribution of structural folding. This finding underscores the importance

Table 4.2. *Logistic regression prediction of revenue decline or high growth*
$(*: p < 0.10; **: p < 0.05; ***: p < 0.01)$.

| Independent variables | Declining revenue (yes=1) | Top quartile revenue growth (yes=1) |
|---|---|---|
| Structural folding | −0.041 | 0.210*** |
| *Intracohesive processes* | | |
| Group size | −0.469*** | −0.144 |
| Capital size of the largest firm | −0.046 | −0.588*** |
| Size difference | 0.061 | −0.159 |
| Financial members | 0.536*** | −0.319 |
| Industry homogeneity | 0.753* | −0.948** |
| *Extracohesive processes* | | |
| Brokerage | −0.026* | −0.006 |
| Bridging ties | 0.019 | −0.040** |
| State-owned proportion | 0.385 | 0.425 |
| Foreign-owned proportion | −0.430 | −0.355 |
| Politicized proportion | 0.984 | −2.904*** |
| Political mix | −0.046 | −0.507 |
| Governing party tie | −0.109 | 0.376* |
| *Controls* | | |
| Group stability from $t − 1$ | −1.685*** | 0.342 |
| Year | −0.064 | −0.037 |
| Group age | 0.033 | 0.162 |
| Newly formed group | −0.480 | −0.049 |
| Labor efficiency (log) | −0.980*** | 1.385*** |
| Capital efficiency (log) | −0.189 | 0.563** |
| Sum of revenues (log) | −0.446* | −0.308 |
| *Industry* | | |
| Energy | −0.181 | −0.884 |
| Mining | 0.741 | 1.880* |
| Chemical | 0.283 | 0.407 |
| Metallurgy | 0.212 | −0.858 |
| Heavy industry | 0.397* | −0.028 |
| Light industry | 0.613*** | −0.274 |
| Wood and textile | 0.524* | −0.039 |
| Food industry | 0.367* | 0.232 |
| Construction | 0.102 | 0.516** |
| Wholesale | 0.616*** | −0.313 |
| Retail | 0.304 | −0.278 |
| Transport | 0.563 | −0.505 |
| Services | 0.431** | 0.142 |
| Constant | 9.850 | −7.637 |
| $N$ | 430 | 430 |
| $-2LL$ | 520.588 | 400.756 |
| $R^2$ | 0.192 | 0.245 |
| % correctly classified | 67.1 | 77.9 |
| $\chi^2 (df)$ | 65.022(33) | 76.679(33) |
| $P - value$ | 0.000 | 0.000 |

of the entrepreneurial generation of ideas through recombination, as opposed to the reliance on imported ideas from other areas of the network.

Playing party politics is a tricky business. While political contacts do not offer protection against decline, only the most narrow, highly targeted strategy has pay-offs. More politicized groups have little chance of achieving high growth. Groups that are overly committed to a party put the trust of their business partners in jeopardy. Groups can only benefit from party politics when they have a tie to a currently governing party.

Structural folding and group stability operate as mirrored opposites. Group stability buffers groups from revenue decline and does not contribute to top quartile revenue growth. Stability in group membership enables reciprocity, solidarity, and mutual assistance to act as a kind of safety net, thereby preventing severe market loss in the group. But trust and improved communication within the group are not assets that stimulate high levels of growth: stability in itself can be a conserving closure.

## 4.6  Structural folds and group stability

Having analyzed how structural folding is related to group performance, we turn to questions of group stability. To define and measure group stability, we refer to Simmel's article "Persistence of Social Groups," where he argues that it is meaningful to speak of group identity, despite shifting membership and low institutionalization, if there is some membership continuity in contiguous stages. We draw on this insight in our analysis of group stability. For the first year of our dataset we use CPM to identify each of the cohesive groups of firms (for example, cohesive group 1 is composed of firms $a, b, c, d, e$; cohesive group 2 is composed of $f, g, h, i, j$; and so on). For the second year we identify the cohesive groups existing at that time (following Simmel's lead, we can call them, $f, b, c, d, e$; $a, g, h, i, j$; $v, w, x, y, z$; and so on). Because network formation is slow at first and the number of groups appearing at the very beginning is only very small, we modify Simmel's scheme. Instead of simply following the first established groups, we identify all the groups that exist from 1987 to 2001. That is, we identify and record, for each year, all of the cohesive groups that exist in that year. By observing the composition of all groups in $t_1$ and those in $t_2$, we can record the proportion of the members of any given group that remained cohesively tied – which is our metric of stability – for all pairs of years.

To measure the stability of groups, we record the flow of members between all groups in adjacent years. A group is completely stable from one year to the next if all members of the group identified in the first year appear together in a group in the next year. At the other extreme, a group dissolves if none of the members at $t_1$ appears in any groups at $t_2$. Between these extremes, a group at $t_1$ can split into segments of various sizes that are present in groups at $t_2$. To measure such intermediate levels of stability, we score the average size of the pieces from $t_1$ that appear in groups at $t_2$, thereby normalizing for the size of the source group.

To test whether structural folding has its own predictive power in the context of competing explanations, we use a multivariate regression model, with group stability as the dependent variable. The independent variables are the same as those that we used in our models of group performance.

We find a negative correlation between intercohesive ties and group stability. The mean correlation across all years is −0.55, ranging from −0.37 to −0.70. This regression model passed Pregibon's link test of model specification; tolerance and variance-inflation factors were within conventional bounds for all independent variables: multicollinearity should not be of concern. Table 4.3 presents the results of our regression models predicting group stability.

We see that multiple membership ties of structural folding decrease group stability. In other words, groups with more structural folds are more likely to break up and, when they do, to break up into smaller fragments. This finding suggests that structural folding stresses the fabric of cohesion. Actors who are ambiguously committed produce destabilizing tensions inside these groups.

Of the intracohesive processes only group size is relevant; larger groups are less stable. Of the various extracohesive linkages only two are relevant to group stability: brokerage and state ownership. Largely as a consequence of the privatization process, state ownership destabilizes groups. This reorganization erodes the stability of groups with state-owned firms.

The number of brokered ties to other groups is significantly correlated with decreased group cohesion, a finding that suggests that brokers adversely impact the structures they exploit. This finding is in line with the idea that the price of brokerage is borne by those who are connected by the broker (Fernandez-Mateo, 2007). In addition to material losses, our findings show another externality of brokerage: the erosion of brokered structures. This structural erosion can eventually diminish the opportunities of brokerage themselves.

Among the control variables, we are not surprised to find that group stability in the preceding year is related to stability in the current year. Group stability has inertia. We also find a positive trend towards greater group stability leading out of the postsocialist period – groups are more stable in the later years. Whereas newly formed groups without a prehistory of cohesion are stable in the first year of their existence, there is a slight disadvantage to old groups.

Groups with higher labor efficiency are more stable, while groups with higher capital efficiency are less stable. A group with a high amount of capital and few employees is much more stable than a group with low capitalization and many employees. This latter type of group was the typical target of reorganization, which

Table 4.3. *Linear regression prediction of group stability (∗ : $p < 0.10$; ∗∗ : $p < 0.05$; ∗ ∗ ∗ : $p < 0.01$).*

| Independent variables | Group stability |
|---|---|
| Structural folding | −0.018∗∗∗ |
| *Intracohesive processes* | |
| Group size | −0.028∗∗ |
| Capital size of the largest firm | −0.009 |
| Size difference | 0.010 |
| Financial members | 0.006 |
| Industry homogeneity | 0.029 |
| *Extracohesive processes* | |
| Brokerage | −0.008∗∗∗ |
| Bridging ties | 0.001 |
| State-owned proportion | −0.078∗ |
| Foreign-owned proportion | 0.061 |
| Politicized proportion | 0.021 |
| Political mix | 0.004 |
| Governing party tie | 0.003 |
| *Controls* | |
| Group stability from $t − 1$ | 0.204∗∗∗ |
| Year | 0.023∗∗∗ |
| Group age | −0.013∗ |
| Newly formed group | 0.127∗∗∗ |
| Labor efficiency (log) | 0.037∗ |
| Capital efficiency (log) | −0.033∗∗ |
| Sum of revenues (log) | 0.004 |
| *Industry* | |
| Energy | −0.041 |
| Mining | 0.086 |
| Chemical | 0.032 |
| Metallurgy | −0.003 |
| Heavy industry | 0.029∗∗ |
| Light industry | 0.045∗∗∗ |
| Wood and textile | 0.068∗∗∗ |
| Food industry | 0.031∗∗ |
| Construction | 0.057∗∗∗ |
| Wholesale | 0.032∗ |
| Retail | 0.019 |
| Transport | −0.009 |
| Services | −0.001 |
| Constant | −1.613∗∗∗ |
| $N$ | 467 |
| *Adj. $R^2$* | 0.472 |
| $F$ *(df)* | 13.671(33) |
| $P − value$ | 0.000 |

disrupted group continuity. In addition, several industry categories are significant, and feature varying levels of group stability when compared with the reference category of agriculture.

To summarize, we found that structural folding is disruptive: groups where membership is not exclusive suffer a loss of stability. Interpenetrating contact between business groups is destructive even beyond what we would expect to happen by chance – groups seem to break down more often if one or more of their members take on multiple affiliations.

## 4.7 Conclusion

Social network analysis has produced a rich array of analytic concepts and powerful methods for studying structural features of economic action. Starting with Granovetter's (1973) idea of weak ties and Burt's (1992) concept of structural holes, a line of research developed these insights into systematic analyses of brokerage opportunities, access to information, and structural constraint. Our work contributes to these efforts by developing the concept of structural folding and then identifying its corresponding topological feature where cohesive group structures fold into each other. In contrast with Burt's image of a structure that bridges or spans a hole, we consider a site where structures fold together.

Behind this difference of imagery lies a further difference in conceptualizing what is transpiring within networks. In the case of brokerage, social networks are channels, a means of transportation, a system of communication. If the flow or movement of information is the critical activity occurring through the conduit or at the contact point across the structural hole, it is the generation of knowledge through recombination that is the critical activity among intercohesive groups. Correspondingly, the electricity metaphor is replaced by more suitable ones from molecular chemistry: we should think of networks as a kind of molecular bonding in which ties connect nodes into larger groupings that represent a new molecular quality, and not merely an extension of circuits to further atoms of the network. Structural folding establishes strong connections between network molecules to generate a more complex material of creative alliances.

Efforts to facilitate innovation should include attention to structural folding. While the idea is simple, attempts at re-engineering atomic group structures into complex molecules often run into serious obstacles. Groups within organizations – project teams, informal workgroups, and management teams – have a natural tendency towards exclusivity. It is easier to think in terms of clear-cut group affiliations, to keep track of who belongs where. Managing a web of overlapping groups can be too complex for effective leadership. It is also emotionally less taxing to maintain clear group loyalties. Identification with a team or division is natural amidst the strong intra-organizational competition within contemporary business firms. Pressures of time management also point towards minimizing group overlaps. A group means scheduling meetings, delivering memos, organizing presentations.

However, there are tools to facilitate structural folds within organizations. An organizational culture that tolerates, or even values multiple affiliations can generate enough openness for group overlaps. Within the context of social movements the idea of open society represents such a cultural element. Promoting a more generalized culture of open association might work within the context of business organizations as well. Beyond culture there are key elements in the organizational environment that might facilitate or block a productive web of group affiliations. Facilities such as office layout, and organizational agoras are obvious conduits for forming structural folds. The availability of information technology tools might also help to maintain multiple affiliations.

Structural folds can be overdone as well. Too many overlapping affiliations will generate too much incoherence and fragmentation, rather than bringing about productive recombinations. Our findings indicate that structural folds de-stabilize networks. Over-promoting group affiliations at the extreme might undermine the organization itself by de-stabilizing the underlying mesh of social network infrastructure.

# 5

# Team formation and performance on nanoHub: a network selection challenge in scientific communities[1]

DREW MARGOLIN, KATHERINE OGNYANOVA, MEIKUAN HUANG, YUN HUANG, AND NOSHIR CONTRACTOR

This study proposes a rationale by which researchers can use network analysis tools to improve policies and guidelines for the formation of teams within a particular community. The study begins by defining voluntary collaborative project teams (VCPTs) as teams where members have semi-autonomous discretion over whether to join or participate. It is argued that, with reference to VCPTs, those making policies or guidelines appear to face both an information disadvantage and possess an information advantage relative to individual team members, stemming from differing positions regarding emergent qualities of teams. While team members have access to team-specific information, policy-makers can use information regarding norms within the community as a whole. It is then argued that network analysis is a useful tool for uncovering these community-wide tendencies. An example analysis is then performed using exponential random graph modeling of a network to uncover the underlying logics of team construction within nanoHub.org, an online community that fosters collaboration amongst nanoscientists. Results suggest that the technique is promising, as a normative tendency which undermines team performance is uncovered.

## 5.1 Introduction

As many industries migrate towards a network-based production implemented by short-term, fluid teams which are permeable, interconnected, and modular (Schilling and Steensma, 2001), the performance of teams where members relatively freely choose their collaborators is becoming increasingly important.

[1] The authors wish to acknowledge the following funding sources that contributed to the development of materials presented in this chapter National Science Foundation Grants: OCI-0904356, IIS-0838564.

Networks in Social Policy Problems, eds. Balázs Vedres and Marco Scotti. Published by Cambridge University Press © Cambridge University Press 2012.

Settings in which semi-autonomous team assembly is particularly prevalent include the development of software in the open source movement, collaboration in large online multiplayer games, scientific co-authorship, and artistic collaborations (Amaral, 2005; Guimerà *et al.*, 2005; Uzzi and Spiro, 2005). Especially noteworthy for governmental funding agencies, research communities, and collaboration platform managers is the growth in the incidence and productivity of science conducted in teams, or "team science" (Fiore, 2008; Stokols *et al.*, 2008; Falk-Krzesinski *et al.*, 2010, 2011; Börner *et al.*, 2010). This growth has made investigations of the mechanisms of effective scientific teamwork and collaboration increasingly relevant. Moreover, recent research has found that many cross-institutional scientific teams fail to achieve their goals; but the few of them that did succeed, did spectacularly (Cummings and Kiesler, 2007; Huang *et al.*, 2010). Therefore, it is important to discover the mechanisms associated with more successful scientific collaborations so that an understanding of these mechanisms can inform policy-making processes for collaboration platform managers and governmental funding agencies. Such information can thus foster more effective scientific team collaborations and improve scientific productivity across communities.

Network research suggests that when organizational entities expand their collaborative activities and diversify their relationships, cohesive subnetworks characterized by multiple, independent logics of organization can emerge (Powell *et al.*, 2005). This self-organized emergence raises a set of important questions for makers of policy and institutions responsible for setting governance structures (Contractor, 1994). Can self-organization be improved? More specifically, can these emergent logics be evaluated from an external position for strengths and weaknesses? Can the introduction of incentives or governance structures guide the development of these logics toward the construction of more cohesive, sustainable, or productive wholes (Provan *et al.*, 2007)?

This study examines these self-organized collaborations in scientific teams. Many contemporary scientific teams can be seen as hybrid-project systems (Schwab and Miner, 2008). In these systems, team membership is neither dominated by organizational control nor fully the result of independent choices made by individuals. In this middle ground, scientists must balance autonomy and institutional constraints when assembling teams to optimize team performance (Huang *et al.*, 2010). As suggested by Schwab and Miner (2008), some hybrid systems tilt more toward the stand-alone end of the continuum, with more freedom of selection on the part of team members, such as research teams with members from different universities. Within this context, the question is how larger institutions might construct and implement policies to provide information, incentives, or guidelines which would improve these semi-autonomous choices.

## 5.2  Voluntary collaborative project teams

The current study focuses on these more freely assembled teams, which we refer to as *voluntary collaborative project teams* (VCPTs). We define voluntary collaborative project teams as those assembled for the purpose of accomplishing a task by a set of individuals who have freedom to choose their collaborators from a large set of potential candidates. VCPTs are what Hackman and Katz (2010) called *self-governing* groups – ones in which members are free to define goals, modify the structure of the team and aspects of its context, and manage their own performance. They go beyond self-governing groups in that individuals have the prerogative to choose who is included in the team.

These teams are distinct from other subjects of small group study. Unlike classic work teams assigned by a superior or an institution, VCPT members select one another and choose to participate on a voluntary basis. Yet unlike emergent groups that lack specific goals, such as social clubs, these teams have specific, finite goals that they intend to accomplish, such as the execution of a particular experiment, the construction of a piece of software, or the publication of an article or chapter. Given this definition, we suggest that performance is the primary goal of voluntary collaborative project teams (Fafchamps *et al.*, 2010), i.e. these teams form because individuals hope that by joining the team they might accomplish something that they might not otherwise be able to accomplish on their own.

In this sense, VCPTs are similar to groups that form to accomplish collective actions, such as the execution of a political protest or the creation of a public good, such as a library or database (Flanagin *et al.*, 2006; Fulk *et al.*, 1996, 2004; Marwell *et al.*, 1988). Yet, unlike in the groups operating according to the classic collective action model, the project cannot be accomplished strictly through the accumulation of sufficient contributions or other a priori specifiable actions. Rather, the project's accomplishment is the result of the group's ongoing process of interaction and collaboration (Poole *et al.*, 1982; Seibold and Meyers, 2007). Thus, choosing the team members who are most likely to help and collaborate well in the successful completion of the project is a complex task. In addition to individual attributes that may suggest competence in tasks required by the project, team members must consider relational factors, such as the ability to communicate with one another, as well as team composition factors, such as the diversity of resources and opportunity for creative discussion (Huang *et al.*, 2010).

Consideration of these features of VCPTs suggests an interesting relationship between VCPT performance and the mechanisms by which VCPTs are assembled. On the one hand, it is in the interest of VCPT members to assemble into teams that achieve high performance. In the absence of institutional requirements which compel participation in particular teams or with particular individuals, team members should be free to volunteer their efforts only to those teams where they believe there will be a sufficient reward, that is, where they believe the team's project will be a successful and valuable one (Fulk *et al.*, 2004). Yet, as individuals subject to the constraints of bounded rationality, it is implausible to suggest that VCPT members choose their teammates perfectly every time (Simon, 1959). Rather, it seems more appropriate to suggest that VCPTs represent individuals' best efforts to construct effective teams. The question for policy-makers is then whether these best efforts can be improved through recommendations from an outside party.

Following this rationale, the goal of this inquiry is to uncover systematic imperfections or biases in individual best efforts. A policy is a systematic intervention in the processes by which a system develops (Ashby, 1960; Provan *et al.*, 2007). To suggest that policy changes that influence processes of VCPT assembly will improve team performance is therefore to suggest that the natural mechanisms of VCPT assembly are very likely to deviate from the ideal in a systematic way.

While there is already some evidence that such systematic biases exist (Guimerà *et al.*, 2005), scholars have not yet examined this proposition formally. In particular, research has not examined whether apparent deviations between observed mechanisms of team assembly and prototypical mechanisms of team assembly can be accurately attributed to systematic factors such as institutional diversity, gender composition, and expertise distribution. The rationale and procedure for conducting such an investigation is the focus of the remainder of this chapter.

## 5.3 Mechanisms associated with successful teams

Research on the assembly of effective teams has identified a number of mechanisms by which successful teams are likely to be formed. Notably, research suggests that teams perform best when they are the appropriate size – large enough to accomplish the necessary tasks, but small enough to minimize coordination costs and communication issues. For example, when new team members are added to an existing team, they are likely to bring three benefits – expertise, information processing ability, and diversity of perspectives. They will also bring a cost – an additional need for coordination (Amaral, 2005).

Teams are also more likely to be successful when they contain individuals with the relevant expertise. Expertise level is one of the most valued resources in scientific and other creative collaboration groups (Huang *et al.*, 2010). For instance, Van Der Vegt *et al.* (2006) examined the effects of expertise diversity on interpersonal helping and found that members will be more committed to and help those seen as having more expertise. Nonetheless, much about the effects of team expertise composition remains to be explained.

In addition, maintaining novelty and freshness in team composition appears to be an important factor influencing team creativity. Guimerà *et al.* (2005) find that teams where members have worked together before are less likely to be creative. However, previous collaboration history among team members can be a double-edged sword, in that prior knowledge reduces uncertainty (Hinds *et al.*, 2000). Huang *et al.* (2010) report that teams composed of individuals from different institutions are likely to provide more innovative products, but this effect is reduced when team members have worked together before. Successful team projects increase the social capital of their creators, which may in turn enhance the popularity of (and the community support for) the new collaborative endeavor (Amaral and Uzzi, 2007).

### 5.4 Translating findings into policy – an information problem

The previous review suggests that there are a variety of mechanisms associated with the assembly of successful teams. However, translating these findings into policy recommendations is not straightforward. While in many cases guidelines and incentives that encourage individuals to assemble teams according to the principles reflected in these findings will be helpful, in other cases they may mislead team members into a new set of problems. The risk of making counter-productive policy stems from the information problem faced by policy-makers: the absence of contextual information regarding a specific team's composition, activities, and internal relationships. Particularly in areas where teams assemble to create novel products or address new problems, recommendations drawn from findings regarding the most effective means by which the previous generations of inventions or solutions were discovered may be misleading. This section describes this problem and elaborates on its application to team assembly.

#### 5.4.1 Emergent local dynamics – the information disadvantage of policy-making

At the core of policy-making is the construction of rules and guidelines for others to follow in pursing their activities. A primary challenge for policy-makers in any field is to justify these prescriptions to the individuals, groups, and institutions that they are guiding. In particular, there is a question of how policy-makers might contribute guidelines for performing an activity to those who are experienced experts in a particular domain. This concern is especially relevant in areas, such as scientific research, where the activities are highly sophisticated, often novel techniques performed by experts in a specialized field.

Both economists and small group researchers have pointed to the disadvantage of guiding local decisions, such as who should work with whom on what, by rules created by a central authority (Hayek, 1945; Poole *et al.*, 1996). The primary argument from these perspectives is that when actions are guided by centralized policy, they take into account a limited set of information. Specifically, prescriptions based on observations of similar situations that have occurred in the past often overlook important information that emerges *in situ*. As Hayek describes:

> The statistics which such a central authority would have to use would have to be arrived at precisely by abstracting from minor differences between the things, by lumping together, as resources of one kind, items which differ as regards location, quality, and other particulars, in a way which may be very significant for the specific decision. It follows from this that central planning based on statistical information by its nature cannot take direct account of these circumstances of time and place, and that the central planner will have to find some way or other in which the decisions depending on them can be left to the "man on the spot." (Hayek, 1945, p. 524)

In small groups, the new information often takes the form of emergent dynamics within a team itself (Poole *et al.*, 1982; Poole and DeSanctis, 1992). Of particular relevance to VCPTs is the emergence of relational factors within the team – the degree to which teammates get along, understand each other, or otherwise find it easy to communicate and cooperate.

In the extent to which these emergent factors influence the appropriate allocation of team resources to achieve team goals, policies imposed from outside and informed by observations of other teams can impede team success. For example, observations which associate team size with task complexity may suggest that individuals facing large tasks form large groups. Such a recommendation ignores the fact that teams may not know the nature of their task until they have been formed and discussion has begun (Weick, 1992). Once formed, however, group members may form relationships, shared understandings, and means of allocating knowledge and responsibility that make it difficult to jettison any particular group member (Hollingshead, 1997, 1998; Wegner, 1987). Thus, groups may become "locked-in" to task definitions which reflect their original size, rather than basing the team's size on a task that is selected by the best judgment of the group (Sydow *et al.*, 2009).

Similarly, the requisite expertise for a task is also contingent on emergent, team-specific factors. Poole and Contractor (2011) argue that assembly mechanisms vary based on the extent to which the teams are engaging in tasks that require exploration of new ideas, exploitation of existing resources, mobilization around a particular issue, bonding and building trust, and swarming in response to an emergency. In addition to factors based on task definition there are also rational factors that need to be considered within the team. The communication of complex knowledge often requires substantial shared understanding that creates absorptive capacity (Clark and Wilkes-Gibbs, 1986; Cohen and Levinthal, 1994). Group members may have an easy time communicating about certain topics or at a certain level of sophistication, but when the task demands the communication of more complex knowledge, communication breaks down (Sorenson *et al.*, 2006). To the extent to which team dialogue and norms of communication emerge through the team's work, it is possible that well configured teams suddenly become ill configured (Cohen and Bacdayan, 1994; Postmes *et al.*, 2000).

While much of the guidance offered by policy will, like team members' own best efforts, be only an approximation for best practices of team assembly, there is reason to believe that these recommendations may contain systematic biases. In particular, when teams are engaged in creative tasks which rely on novel, expert knowledge, repeating the team assembly processes that have led to success in the past may – like repetition of effective collaborations in the past – lead to outputs that are similar to what has been produced in the past (Guimerà *et al.*, 2005). Thus, as consistency with previously successful practices is enforced, novelty and diversity may be lost (Sydow *et al.*, 2009). For example, Gerstner (2002) suggests that IBM's stagnation emerged in part because it followed habits and routines that had brought it success.

Since policy-making is constrained to provide rules based on prior observations, these arguments suggest a limited role for the guidance of team formation based on findings from team assembly. The next section presents a counter-argument to the claim.

### 5.4.2 Emergent collective dynamics – the information advantage of policy-making

In the previous section it was argued that policy-makers are at an information disadvantage vis-à-vis team members in assessing appropriate team composition. Specifically, it was suggested that policy-makers have access to information that is more homogeneous than members of individual teams because these team members can combine best practices learned form historical observations with specific observations of emergent team dynamics. In this section, it is argued that team members are also burdened by their own yoke of conformity, and that policy-makers may be able to alleviate this burden through analyzing information that they are uniquely privy to. This information is in regard to a different set of emergent qualities, what can be called *emergent community norms*, rules or habits of team formation which are shared by a substantial portion of teams within a particular community (Meyer and Rowan, 1991 [1977]; Nelson and Winter, 1983).

As described above, team members can develop rules and norms that are specific to each team through the course of its operation (Poole and DeSanctis, 1992; Postmes *et al.*, 2000). Yet in many cases, team members are also members of higher-level institutions or communities (Lammers and Barbour, 2006). These higher-level institutions can also develop emergent norms which guide or constrain the behavior of team members (Monge and Contractor, 2003). For example, Huffaker (2010) reports a variety of tendencies that are shared across discussion topics and groups in an online community. Individuals are not likely to be aware of the fact that their individual sub-communities share common patterns with one another.

These norms can be helpful in guiding individuals to make better decisions. The generally accepted rule that researchers demonstrate competence through the achievement of a PhD is likely to be an effective rule. As research in neo-institutional theory argues, however, there is no guarantee that emergent norms correspond to performance enhancements (Meyer and Rowan, 1991 [1977]). That is, individuals or groups may consistently engage in a variety of routines, behaviors, or use logics of evaluation and inference simply because others also do so, or because these have been the rules that have been used in the past rather than because adherence to these rules provides any particular advantage.

In particular, institutional scholars point to the fact that consistent behaviors within a field or institutional setting often reflect the influence of information limitations (DiMaggio and Powell, 1983), rules for legitimate action (Hannan and Freeman, 1977; Meyer and Rowan, 1991 [1977]), as well as policy prescriptions (George *et al.*, 2006). When information regarding effective practices is limited, groups often engage in *follow-the-leader* behavior in which they imitate those groups which appear to be most effective (DiMaggio and Powell, 1983). In essence, these groups over-rule or over-ride their local, contextual information in favor of whatever the best performing group is doing. While this might bring an advantage, in many cases it does not. A single leader is a small sample size and many contextual factors which imitation ignores may be relevant. For example, Huffaker (2010) reports that group members are likely to imitate the language patterns of individuals that post frequently and respond regularly to others. Huffaker's results also show that the linguistic diversity of group leaders can foster further discussion within the team. It is thus possible that groups where certain individuals assert leadership through frequent posts and replies, but use a limited, homogenous vocabulary, may suppress further group discussion even when such discussion would be productive.

*Legitimacy* refers to degree to which a rule or practice is considered valid and worthy of resource investment by a community (Hannan and Freeman, 1977; Pfeffer and Salancik, 1978). When groups engage in legitimate practices, they are likely to continue to receive resources even if their actions fail (Baum and Oliver, 1991). Groups can also be guided by official rules and policy which do not correspond to performance (George *et al.*, 2006).

Research in these areas suggests that individuals, groups, and organizations are likely to continue to follow these practices, called *myths* by Meyer and Rowan (1991 [1977]), for at least two reasons. Firstly, there is evidence that individuals are not even aware that they follow these rules (Cohen and Bacdayan, 1994; Nonaka and Takeuchi, 1995; Zucker, 1991 [1977]). Secondly, the identification of myths requires information regarding the typical behavior of groups within a community. While in some cases this information is readily available, it is often distributed widely through the community. In particular, in cases where individual groups interact only with a small number of others, it may be difficult to observe broader tendencies within the community as a whole. It is unlikely, for example, that Google Groups participants are aware of the larger patterns in leadership and conversation that can only be observed through an analysis of thousands of posts (Lazer *et al.*, 2009). Such global knowledge is also unlikely in cases where groups share a common platform or technology but communicate on a limited basis. In particular, actions or decisions that appear novel and objective within a local area of a community may, in fact, be typical and normative within the community as a whole (Burt, 1987; Strang and Meyer, 1993).

The existence and systematic influence of these community-level myths or rules offer an avenue through which policy-makers may make a contribution to team assembly practices. In particular, policy-makers are a potential source of information regarding the biases that individuals may not be aware of. In many cases, policy-makers have access to information regarding the community as a whole. Particularly in digital communities, administrators or other policy-makers can observe the behavior of the whole community in a way that individual members cannot (Lazer *et al.*, 2009). As such, they can observe individuals embedded in a multidimensional network where some of the nodes are individuals, but others are technological artifacts such as documents and databases, as well as conceptual artifacts such as keywords and tags (Contractor *et al.*, 2011). More importantly, policy-makers have incentives to access and analyze this information that is not shared by individual teams. As beneficiaries of the aggregate of performance across all teams, policy-makers gain from any insight that can be gleaned from the analysis of this community-wide information (Easley and Kleinberg, 2010; Lazer and Friedman, 2007). More specifically, policy-makers stand to gain from any insight that can be used to improve *any* team over which they have policy-making authority. Individual teams do not necessarily share in this reward. Individual teams only benefit if analysis of the aggregate, community-wide information improves *their own* performance. Thus, the analysis of community level norms as potential myths is likely to fall prey to collective action failures (Fulk *et al.*, 1996, 2004).

## 5.5  Using network analysis to assist policy
### 5.5.1  Network analysis and community norms

Network analysis offers a powerful tool for identifying emergent norms within a community. Collective network structures have been shown to be important predictors of a variety of individual behaviors through processes of diffusion (Rogers, 2003). That is, the position of an individual in a network is often an important predictor of that individual's behavior. Yet individuals generally lack information regarding their position (Krackhardt, 1987). Networks are thus an important, hidden influence on behavior. The individual, or individual group, may experience network influences as changes in the salience or relevance of particular cues or values (Strang and Meyer, 1993; Strang and Soule, 1998). Without access to information about the network as a whole, however, the degree to which these changes are recognized as emergent norms is likely to be limited. More specifically, what individuals experience as novel may, through network analysis, be revealed to be normative (Coleman *et al.*, 1957; Burt, 1987).

For example, Barabási and Albert (1999) have shown that the citation of scientific papers suggests a normative logic called *preferential attachment*. Preferential attachment is a form of institutional imitation. According to preferential attachment, papers that receive many citations are more likely to be cited by new papers, strictly because they have received other citations in the past. The result of preferential attachment is a highly skewed distribution of citations where few papers receive a large number of citations and most papers receive very few. As Evans (2008) shows, it is quite likely that this habit does nothing to increase the quality of scientific research and may even harm it by limiting the diversity of ideas.

In spite of its potential for limiting scientific performance, evidence suggests that preferential attachment is an emergent rule rather than one that is consciously evaluated and chosen by researchers (Barabási, 2003; Powell *et al.*, 2005). As Barabási remarks, the mechanism appears to have operated in a variety of fields both in science and outside of science without anyone's knowledge. It was only through the analysis of large collections of citations that this norm was uncovered.

### 5.5.2  Network signatures of emergent norms of team assembly

Based on the preceding arguments, we suggest that policy-makers can use network analysis to identify community norms of team assembly. This section will explain the rationale by which these analyses can be used to make new rules and guidelines for the formation of teams. An example of such an analysis using data from nanoHub.org is then provided.

The preceding arguments suggest that the logics by which teams are assembled can be categorized in one of three general categories. Firstly, there are logics of team assembly which are typical in a community because they improve performance. These logics represent knowledge the community has gained. These may include some of the logics that have been discovered by team assembly researchers, and they may also include logics which have been uncovered within the community but are not yet known to team assembly researchers. Given that VCPTs involve voluntary contributions of team members' time and resources in order to accomplish goals, it is likely that many of these performance enhancing logics are present.

H1: For some logics of team assembly that are typical in a community, conformity to this logic will be associated with positive performance.

At the same time, the arguments from institutional theory presented above suggest that some logics may not have any performance benefit. In particular, these arguments suggest that logics which are difficult to observe from individual positions, such as preferential attachment, are not likely to be associated with performance. This is due to the fact that these logics may take hold without being evaluated by individuals for their performance benefit. Thus:

H2: For some emergent logics of team assembly that are typical in a community, conformity to this logic will be associated with negative performance.

While there is good reason to believe that preferential attachment will be such a logic, other network logics, such as imitation through structural equivalence (Burt, 1987) and repeat collaboration (Guimerà *et al.*, 2005) are also candidates. A third possibility is that team assembly logics are idiosyncratic or experimental, that is, that they are not typical within a community. These can be thought of as undiscovered by researchers or un-diffused within the community. These experimental logics may be performance enhancing, or they may be performance undermining. Policy-makers may be interested in finding the experimental logics that are performance enhancing. Thus we ask:

RQ1: Are any atypical, experimental logics of team assembly performance enhancing?

## 5.6 Method

### *Sample*

nanoHub is a major NSF-funded cyberinfrastructure-enabled collaboration environment hosted at Purdue University. It allows scientists and engineers to develop software tools and perform simulations for nanotechnology research. The online platform is a global resource for nanoscience and technology, created by the NSF-funded Network for Computational Nanotechnology. The software tools on nanoHub are hosted and run on computer clusters equivalent to 100 personal computers. Researchers around the world can access the platform to run simulations and perform experiments. This can be done through a regular Web browser run on personal computers or even a cell phone. Thus nanoHub provides researchers with the resources to do more scientific work without having to spend the time and money to build their own simulation environment.

nanoHub facilitates software tool collaboration in a number of ways. It provides a platform for software simulation, reducing development costs and creating a global scientific community of software developers and users working in the area of nanotechnology. Through shared virtual workspaces, nanoHub allows groups of users to collaborate, share results, and do collective troubleshooting. nanoHub facilitates the formation of virtual teams of scholars working on software development projects.

An explicit goal of the nanoHub online platform is to foster interorganizational and international collaboration. Combined with the ease of use and virtual nature of the workflow, we expect this to alleviate constraints based on structural opportunities for social contact (described in Ruef *et al.*, 2003). At least in theory, all who are interested in launching a project should have the means to contact the desired potential team members through the online service. This idea is, furthermore, in keeping with the definition of voluntary collaborative project teams that this chapter employs.

The sample used in this study consisted of 124 teams (size ranging from 2 to 8) with a total of 170 researchers who collaborated on the development of software tools published on the nanoHub. The researchers were distributed across 32 organizations (96% were universities) located in 10 countries. Ninety percent of the researchers were in the USA, with the rest located in Australia, Canada, China, Britain, Italy, Malaysia, Spain, and South Korea.

### 5.6.1 Inferring community logics

To infer the logics of team assembly within the nanoHub community we employ an Exponential Random Graph Model (ERGM) for bipartite networks. Through Monte Carlo Maximum Likelihood Estimates (MCMLE), the exponential random graph modeling approach identifies parameters which, if used as guides for the construction of the network, would make the observed network likely (Robins *et al.*, 2007). If the assembly mechanisms of a large number of teams can be explained by the operation of a small number of parameters as guides, this suggests that these parameters are close approximations to the tendencies by which team members come to work together.

Importantly, this technique involves an analysis of network outcomes, not individual intentions. The parameters that guide network construction may not be the rules by which individuals choose to form teams. Rather, these parameters describe the tendencies of team formation which emerge through the combination of individual motivations, inferences, and their interaction with contextual information. As such, ERGMs provide an excellent tool for identifying the emergent logics in a system or community.

For the purposes of this study, we constructed a two-mode network of 170 nanoHub researchers and their affiliations with the 124 tool projects on the platform. Using the BPNet software (Wang *et al.*, 2009) we created a model of the observed network which included the following potential underlying logics of team assembly.

### Individual attributes

The knowledge, skills, and other individual characteristics of researchers affect their chances of being on a team. Previous research has shown, for instance, that gender, status, and expertise are important in the selection process (Hinds *et al.*, 2000). Factors like individual and aggregate team member expertise are also expected to be important for productivity (Balthazard *et al.*, 2004).

Additionally, particular individual characteristics specific to the nanoHub platform may play a role in team formation. Researchers belonging to the Network for Computational Nanotechnology (NCN), the founding organization of nanoHub, can be expected to play a more active role on the platform, starting and joining a large number of development teams.

The model proposed here looks into the impact of the following characteristics:

- **Gender**: nanoHub teams are mostly composed of male researchers. While there are a small number of female tool developers on nanoHub ($N = 9$); those women researchers seem to play an active role: out of 124 teams, 30 have at least one female member. The gender parameter included in our model addresses the possibility that women are in general more likely to join teams.
- **Expertise**: knowledge and expertise are important qualities of a team member. We expect individuals with high observable level of expertise to be more likely to appear on many teams. For the purposes of this study, expertise is assessed based on each researcher's number of academic publications. The scholars who ranked in the top 25% for number of published papers were considered experts ($N = 49$).

- **NCN**: as members of the Network for Computational Nanotechnology are more invested in the nanoHub platform, they are likely to be particularly active members of the software development community. We expect individuals belonging to NCN to be more likely to participate in teams.

### *Structural parameters*

In addition to individual attributes, the bipartite exponential random graph model developed here involves a number of structural properties of the researcher–team network. Parameters for alternating $k$ stars are used to control for team popularity effects (the presence of a few large teams that many people have decided to join) and individual member attractiveness or extroversion (researchers likely to create, be invited onto, or volunteer to join a large number of projects). Additional structural parameters control for network density (number of links) and the propensity of researchers to work with the same collaborators on multiple teams (alternating $k$ two-paths).

### *5.6.2 Dependent/performance variables*

As community logics are assessed in terms of their impact on project outcome, we consider a number of measures related to the popularity, attractiveness, and quality of the software tools produced by each team in our sample. Since nanoHub software tools run on the platform's own computer clusters, nanoHub has accurate logs describing user ratings, tool tagging, and software citations. We take advantage of this rich source of information to construct a composite measure of team performance. The following five indicators are used for this purpose, as their combination captures both the academic impact and the user satisfaction with a particular software project.

(1) **Tool rating**: users of the nanoHub platform can rate the quality of each software tool. The scores given to the projects in our sample serve as an indication of the usefulness and quality of the software.

(2) **Number of times rated**: since the rating score of each tool is computed as an average, the number of people who opted to rate the project is another important measure. Popular, well-used tools are more likely to be rated by a large number of users.

(3) **Number of times cited**: nanoHub teams produce software that is used and cited in the academic community. As academic citations are often used to assess the originality, quality, and influence of the work produced by university research groups (Wuchty *et al.*, 2007), the number of citations (collected from published academic papers) is a good way to assess the impact of team output.

(4) **Number of times tagged**: both users and team members on nanoHub can tag software tools in order to make them easier to discover and categorize. Popular projects, as well as complex or interdisciplinary instruments, are likely to have a large number of tags applied to them by community members.

(5) **Number of taggers**: as tags can be generated by the creators of a software tool themselves, another relevant measure here is the number of nanoHub members who added a tag description to each project. Software tagged by a large number of people can be expected to be both popular and complex in nature.

The outcome variable used for hypothesis testing in the study is therefore computed taking into account all of the aspects mentioned above. Performance is operationalized as the sum of the standardized scores of five indicators: rating, times rated, number of citations, tags, and taggers.

### 5.6.3 Hypothesis testing

Hypotheses 1 and 2 are tested using the following procedure. Firstly, an ERGM is fitted to the network of teams. This model reveals the dominant parameters guiding team assembly in nanoHub. Next, these parameters are used to identify the degree to which teams conform to these dominant logics. As described above, some teams will conform to dominant logics whereas others will be idiosyncratic or experimental. The degree of conformity to a particular logic is then included as a predictor in a regression on the various team performance measures.

#### Calculating conformity

The degree to which a team conforms to a logic can be calculated in a number of ways. For individual logics, those that relate to structural features or attributes of particular individuals, the count of individuals included on the team that display those features is used. For example, if the ERGM suggests that teams favor the inclusion of members with a certain level of expertise, then the conformity of a team is based on the number of members with that level of expertise that it contains.

For more complex, collective logics, specific measurements related to the level of the parameter can be devised. For example, if a parameter showing that teams of a certain critical size tend to become very large, then a team's size relative to this threshold can be used. For the purposes of this study, five underlying logics were selected and explored in terms of their importance for team assembly and their impact on performance. Those are:

- **Gender-related**: (1) preference for selecting females on a team and (2) impact of number of female team members on performance.

- **Expertise-driven**: (1) preference for selecting experts on a team and (2) impact of number of experts on team outcome.
- **NCN-driven: nanoHub specific**: (1) tendency to include NCN members on a team and (2) impact of number of NCN members on performance.
- **Star-related**: (1) the propensity of a few popular/extraverted individuals to participate in a large number of teams and (2) the effect of the number of star members on team outcome. For the purpose of this study, star members are those whose number of team affiliations is more than one standard deviation larger than the mean for individuals in the sample ($M = 2.32, SD = 2.92$).
- **Team size**: team size was included as a control variable in both models. The number of members has a complex role in patterns of team assembly and performance. While more detailed inquiry into the dynamics of different team sizes is possible, for the purposes of this study the aspects of size controlled for were (1) propensity of select teams to attract a large number of members and (2) effect of team size on project outcome.

### *Regression on performance*

After conformity measures were calculated for each logic, Hypotheses 1 and 2 were tested with a regression of team performance on these measures. According to Hypothesis 1, for at least one dominant logic, conformity to the logic will be associated with improved performance. In other words, teams with a higher degree of conformity to the logic will show a higher performance. Thus, the beta coefficient for conformity for at least one dominant logic will be positive. According to Hypothesis 2, at least one dominant logic will be associated with depressed performance. Thus, the beta coefficient for conformity for at least one dominant logic will be negative. In other words, teams with a higher degree of conformity to the logic will show lower performance.

## 5.7 Results

### 5.7.1 *Identifying community logics*

To test conformity to community logics, an exponential random graph model of the observed bipartite network was generated. A BPNet estimation of the goodness of fit confirmed that the model was satisfactory, with all of the parameters having t-statistics lower than 0.1 in absolute value.

The converged model (Table 5.1) suggested that, on a community level, there is a high degree of conformity with the logic that increases the probability of NCN members joining teams ($Estimate = 4.9$, $SE = 1.47$). No such patterns can be observed for females ($Estimate = 0.18$, $SE = 0.27$) and experts ($Estimate = 0.11$, $SE = 0.13$). A significant negative team star parameter ($Estimate = -0.86$, $SE = 0.28$)

Table 5.1. *Team assembly ERG model – community logics (* $* = t < 0.1$, *ratio of estimate to standard error* $> 2$).

| Parameter | Effect estimate | Stdandard error | t-ratio | Significant |
|---|---|---|---|---|
| Density | −2.737 | 0.476 | −0.003 | * |
| Member stars | 0.162 | 0.268 | −0.010 | - |
| Team stars | −0.860 | 0.275 | 0.004 | * |
| Concurrent collaboration | −0.205 | 0.118 | −0.026 | - |
| Experts on teams | 0.107 | 0.133 | 0.008 | - |
| Females on teams | 0.177 | 0.267 | 0.053 | - |
| NCN members on teams | 4.903 | 1.471 | −0.035 | * |

Table 5.2. *Regression model: community logics – impact on performance; the relative strength of individual predictors (* $*p < 0.05$).

| Predictors | Standardized regression coefficients $\beta$ | Zero-order correlation with performance | Partial correlation with performance |
|---|---|---|---|
| Team size | 0.176 | 0.397* | 0.161 |
| Number of females | 0.183* | 0.296* | 0.182* |
| Number of experts | 0.236* | 0.401* | 0.207* |
| Number of stars | 0.424* | 0.341* | 0.260* |
| Number of NCN | −0.457* | 0.165* | −0.286* |

indicates that large sized teams are improbable and the variance in team sizes is not very big (Robins *et al.*, 2006).

### 5.7.2 Performance regression

Multiple regression analysis was conducted to evaluate the relationships of team performance on conformity measures including the team size and the number of females, experts, stars, and NCN members. The linear combination of those five factors was significantly related to team performance, $R^2 = 0.29$, $adjusted\ R^2 = 0.26$, $F(5, 118) = 9.54$, $p < 0.001$. Even though the explanatory power of the regression model could be improved by removing some of the predictors, all of them were retained. This was done to allow for a look at the team assembly logics listed in the previous section and their effect on project outcomes. The relative strength of individual predictors is presented in Table 5.2.

Table 5.3. *Correlations between predictor variables (∗p < 0.05, ∗∗p < 0.001).*

| Predictors | Team size | Number of females | Number of experts | Number of stars | Number of NCN |
|---|---|---|---|---|---|
| Team size | 1 | 0.302∗ | 0.569∗∗ | 0.494∗∗ | 0.389∗∗ |
| Number of females | | 1 | 0.385∗∗ | 0.452∗∗ | 0.486∗∗ |
| Number of experts | | | 1 | 0.519∗∗ | 0.494∗∗ |
| Number of stars | | | | 1 | 0.821∗∗ |
| Number of NCN | | | | | 1 |

Four of the five predictors in the regression model were statistically significant ($p < 0.05$), team size being the only exception ($p = 0.79$). Three of the indicators – number of females, experts, and stars – had a significant positive zero-order and partial correlations with the performance variable. Number of NCN members had a significant positive zero-order correlation with team outcome, $r(124) = 0.165$, $p < 0.05$, but a negative partial correlation, $r(124) = -0.286$, $p < 0.05$, when controlling for all other indicators. The strong correlation between the number of stars and the number of NCN members may account for this ($r(124) = 0.891$, $p < 0.001$, see Table 5.3). This information suggests that NCN members are helpful to performance, because most NCN members that appear on teams are also stars. Controlling for stars, however, NCN members place a drag on performance.

## 5.8 Discussion
### 5.8.1 Review of findings

Results of the hypothesis tests, as summarized in Table 5.4, suggest that this method is promising. In particular, Hypothesis 2 is supported. That is, one logic, the inclusion of NCN members, is typical in the community but is performance suppressing. Evidence regarding Hypothesis 1 and the research question is a bit more complex. Hypothesis 1 is technically not supported, as no logic that is typical of the community is associated with increased performance. The three logics which are associated with improved performance – the inclusion of women, the inclusion of experts, and the inclusion of "stars" – do, however, each have a positive effect estimate in the ERGM. As the estimates are not statistically significant, the logics cannot be deemed typical of the community. They do, however, appear to be somewhat commonplace.

Thus, while Hypothesis 2 receives stronger support, the evidence from an analysis of the nanoHub community as a whole does suggest that nanoHub participants are using a combination of effective and ineffective team assembly mechanisms. Many members appear to be following useful logics by including women, experts,

Table 5.4. *Hypothesis testing.*

| Team attribute | Typical community logic? (based on ERG model) | Effect on performance? (based on regression model) | Hypothesis testing (logics and outcomes) |
|---|---|---|---|
| Team size | (Control variable) | N/A | N/A |
| Gender | NO | POSITIVE | RQ1 Tentative Yes |
| Experts | NO | POSITIVE | RQ1 Tentative Yes |
| Stars | NO | POSITIVE | RQ1 Tentative Yes |
| NCN members | YES | NEGATIVE | Supports H2 |

and stars, though these habits are not dominant. Some of these benefits are suppressed, however, by the general tendency for nanoHub teams to include members affiliated with NCN institutions. This tendency is likely to reflect the degree to which NCN members are more familiar with the platform or with one another. It may also reflect policies or pressures from their institutions which encourage their participation.

Yet whatever the cause, the favoring of NCN members as teammates appears to be weakening performance. That is, individuals would be better served by searching more broadly within the nanoHub community for potential collaborators, potentially increasing both the diversity of interactions and the diversity of approaches to working with the platform.

It is also possible that NCN membership is interacting with team size to suppress team performance. As articulated in the theory section, once formed, teams may experience a lock-in to their size. The evidence from the ERGM further suggests that, in nanoHub, teams are on average smaller than could be expected by chance. Taken with the significance of the NCN parameter, this suggests that nanoHub users may be prematurely "filling" the limited roles in teams with fellow NCN members when, in fact, members of non-NCN institutions may make better collaborators.

Returning to the other helpful logics that are not typical, experts and stars present an interesting case. It is not surprising that the usefulness and scarcity of experts is a significant predictor of performance. Presumably, individuals that have a strong record of publication in the field of nanotechnology have access to knowledge and skills that are useful to other teams in the production of new nanotechnology tools and ideas. Interestingly, however, this effect does not explain away the additional contribution of stars. That is, individuals who are not experts, in terms of publication, but are stars, meaning they work with many individuals in nanoHub, also provide performance benefits to teams. It is possible that these individuals are "up-and-coming" experts who have not yet built a publication record and are thus making a contribution on a similar basis to that made by established experts. It is also possible, however, that these individuals are making a nanoHub-specific contribution. That is, these individuals may possess knowledge regarding the effective construction of new tools or other products within the nanoHub platform.

The significant, positive contribution of women to teams within nanoHub is also interesting. One possible explanation is that women are simply superior team players. That is, because of their ability to facilitate communication and teamwork (Metcalfe and Linstead, 2003), teams with women are more successful. Another explanation is that this finding reflects the residual evidence of gender discrimination in science. More specifically, because women find it more difficult to establish themselves in sciences such as nanotechnology, those women that do make it in the field tend to be disproportionately talented.

### 5.8.2 Policy implications

These arguments suggest that nanoHub administrators may wish to consider the following policies or guidelines to improve team performance within nanoHub:

(1) Set a quota of NCN members per team, based on team size. For example, teams could be required or encouraged to use no more than two NCN members unless they were at least five members in size. The evidence suggests that a quota would be helpful for two reasons. Firstly, a quota will encourage the inclusion of non-NCN members in teams, potentially improving performance. Secondly, such a rule will not force teams to necessarily abandon helpful NCN partners, as teams can also increase diversity by increasing team size.

(2) Make information and opportunities to collaborate with experts and stars more available. While it is possible that experts and stars are at their personal capacities for collaboration, it is also possible that teams do not include a sufficient number of experts and stars, because many teams do not know which experts and stars are available to work within nanoHub. Asking experts or stars who are willing to take on more projects to provide a form of contact or invitation might be helpful.

(3) Provide gender-disguising searches for collaborators. More work needs to be done in this area, but the preliminary evidence suggests that women were not statistically more likely to be on teams than men, given the number of women in the sample, but that team performance significantly improved performance when women were included as members of a team. This suggests that there may be some discrimination based on gender. In particular, it is possible that the women in the sample represent better overall performers than the men. In this case, the observed effects show a bias towards choosing men. That is, a

woman has to be of higher quality to be chosen at the same rate as a man, but once she is chosen, this higher quality is reflected in better team performance. One way to reduce the traditional unfair and performance limiting bias towards females would be to provide information about potential team members that is stripped of gender markers, such as photos and first names. The small, non-significant parameter favoring the inclusion of women may also be due to the small number of women in the sample. If more women were present, it is possible the effect would have been significant and positive, showing that, as predicted by H1, a typical logic of team assembly (choosing women as team members) is associated with higher team performance.

### 5.8.3  Limitations and further research

The research reported here is preliminary in many respects. Firstly, the data reflect only a sample of 124 teams taken from a particular time window. A more thorough analysis would include many different time slices and identify parameters whose significance was robust within the community. Secondly, further interrogation of relevant ERGM parameters should be considered. The parameters chosen here were largely those based on previous findings in team assembly research. However, there is no reason to believe that nanoHub members have not uncovered a variety of distinct logics, some performance enhancing, some performance neutral or undermining, that are unique to this platform or to online communities in general.

Thirdly, it is important to consider the influence of selection effects (Heckman, 1976, 1979). In particular, the data here are extant teams – those which were not only assembled, but performed sufficiently well to produce an observable output. However, to the extent to which the production of this output is correlated with variables that influence team performance, it is likely that teams assembled according to ineffective logics simply were not observed. Thus, there may be other logics which might support H2. Conversely, as Heckman (1979) shows, the exclusion of selection effects can suppress significant effects in the final regression. That is, many teams may be assembled by typical logics which provide them with a requisite level of performance enhancing resources. Having achieved this requisite level, these teams vary widely in their performance. Having failed to achieve this level, teams are not observed. In this case, the beta coefficient associated with this logic would be non-significant even though the logic is helpful to performance.

Further research should address these limitations. The method proposed here can also be extended to other communities where team assembly is likely to occur in a shared technological or normative environment. Such areas include scientific collaboration within particular disciplines or sub-fields, collaboration in particular industries, and voluntary task groups that form in various online communities.

# Part II

Influence, capture, corruption: networks perspectives on policy institutions

# 6

## Modes of coordination of collective action: what actors in policy-making?

MARIO DIANI

Researchers interested in exploring the role played, within policy networks, by organizations representing collective and/or public interests are faced with an array of options which is as impressive as potentially confusing. They may concentrate on the organizations considered most influential in a certain policy domain (e.g. Laumann and Knoke, 1987), or they may look at the entire field of organizations operating on a certain issue or set of linked issues (e.g. Diani, 1995); they may focus on organizations that are closest to the model of the social movement (e.g. Amenta *et al.*, 1992) or, instead, pay attention to public interest groups in the broadest sense, sometimes without even acknowledging any meaningful difference between social movement organizations and interest groups (e.g. Burstein and Linton, 2002).

While each of these strategies may offer valuable insights, it is important to be explicit regarding the types of actors whose role in policy-making we want to explore, and on the relations that they entertain to other actors with similar agendas. Although their internal structure may substantially differ, organizations representing public interests are not usually strong enough to be able to operate entirely on their own. In most cases, they are involved in some kind of alliance and cooperation with cognate organizations. Even if they operate mostly independently, they are nonetheless located within broader organizational fields. This has some important consequences for the analysis of policy networks. Firstly, studies focusing on the most conspicuous and influential organizations in a given policy domain cannot take for granted that they will be representative of the organizational field taken as a whole. This may result in serious problems of legitimacy when it comes to the implementation of certain policy decisions. Secondly, from the opposite perspective, too inclusive studies may end up drawing conclusions on the role

Networks in Social Policy Problems, eds. Balázs Vedres and Marco Scotti. Published by Cambridge University Press © Cambridge University Press 2012.

of "social movements" in policy-making, when they are actually talking about fairly classic interest groups. More in general, we risk making inferences about the part played by large categories ("environmentalists," "ethnic organizations," or the like) in policy-making, without recognizing the huge internal differentiation within them, nor – most importantly – the variable relations that may connect them to each other.

This chapter proposes a framework that enables us to recognize the complexity of organizational fields and the heterogeneity of the relational patterns operating within them. It borrows from organizational theory the expression "modes of coordination" to identify the different forms through which the representation of public and collective interests can be conducted. Moving from the most ambiguous and contested concept, that of "social movement," it identifies three other modes of coordination, defined as "organizational," "coalitional," and "subcultural/communitarian." They originate from different combinations of the responses that organizations provide to general problems of resource allocation and boundary definition. The typology is illustrated with empirical evidence from a study of Milanese environmentalism in the mid 1980s. The chapter concludes with some indications about the implications of this particular perspective for the analysis of policy networks.

## 6.1 Modes of coordination within organizational fields

The organizations promoting collective action and advocacy on public and/or collective issues may be conveniently looked at as part of broader "organizational fields." One of the main problems associated with the study of advocacy organizations' fields[1] lies with analysts often focusing on the properties of the actors operating within them rather than on the patterns of relations connecting them to each other. For example, social movements are frequently analyzed as the collection of organizations interested in a certain set of issues (Dalton, 1994; Andrews and Edwards, 2005), or as sets of like-minded individuals sympathizing with a certain cause or, if we are to impose stricter criteria, willing to adopt certain (contentious) forms of action (Dalton, 2008; Norris, 2003). While all these approaches are useful and relevant, they all reflect a reductionist view of structure as an aggregate of actors sharing certain traits, rather than as a distinct pattern of relations between those elements (Kontopoulos, 1993, Chapter 1). Accordingly, the "structure" of a collectivity (whether labeled a "movement," a "field," or something else) is ultimately given by means and percentages, i.e. by the distribution of the traits commonly associated

---

[1] Although in general the concept of the field includes all the organizations that represent a recognized area of institutional life, including public agencies, business actors, etc. (Powell and DiMaggio 1991, pp. 64–65), in the context of this chapter and for the sake of simplicity when referring to "fields," I will restrict its meaning to sets of voluntary organizations.

with it: for example, the preferred styles of participation among people identifying with a given cause, or the issue priorities, the strategies, or the organizational traits of the organizations mobilizing on certain issues.

While this approach generates accurate descriptions of population traits, it tends to leave out how such traits combine in specific relational patterns. It indeed makes a lot of difference whether the organizations interested in certain issues collaborate, mutually supporting their respective initiatives, and blending them in broader agendas, or whether they do work independently, trying to secure themselves a specific niche. For example, in the case of environmentalism, comparative evidence suggests that animal rights issues may or may not be linked to more classic political or conservationist agendas, depending on the traditions of different countries (Rootes, 2003); only in the former case would it make sense to identify animal rights groups as a major player in the environmental policy network.

The lesson from these examples is pretty clear: it is not just the properties of individual organizations that matter, but also the networks that connect them to each other. Recognizing this requires some re-adjusting of our perspective, in particular, of the way we look at networks. While networks are mostly treated as preconditions for collaborative action by analysts of both social movements (McAdam, 2003; Tilly and Tarrow, 2007, p. 114) and organizations (Uzzi, 1997; Campbell, 2005, pp. 61–62), a different if not incompatible approach has also developed, stressing that networks are also generated through collaborative action (Gould, 1991; Diani, 1995; Gulati and Gargiulo, 1999; Ansell, 2003; Diani and Bison, 2004).

In this view, social networks originate from repeated discrete decisions that a variety of groups and organizations make regarding their partners in projects and campaigns. They also stem from similarly heterogeneous decisions that individual activists make in relation to their multiple memberships, their involvement in collective activities, their personal ties to fellow activists. Not all of these decisions combine to reflect some pattern, nor do they all display continuity over time; most actually do not. When they do, however, they create an informal yet relatively stable model of social organization. In such a model, that bears more than passing analogies to models of network organization (Podolny and Page, 1998) and network governance (Sorensen and Torfing, 2007), a plurality of actors engage in sustained cooperative exchanges in pursuit of a common goal, while maintaining their autonomy, and negotiating a balance between specific organizational identities and loyalties addressed to broader collectivities.

At the same time, the current spread of the network metaphor to characterize the operation of complex organizational populations must not generate the false assumption that network forms of organization be a universal property of organizational fields. To the contrary, within specific organizational fields we may find different logics of collective action at work, ranging from the highly informal and decentralized to the highly formalized. There are, in other words, different "modes of coordination" of collective action at work.

The concept of "mode of coordination" (for earlier uses see, among others, Camerer and Knez, 1996; Mayntz, 2003; Thevenot, 2001) is ultimately rooted in the tradition of typologies of forms of social regulation stretching back at least as far as Karl Polanyi (or, one could argue, the sociological classics). Much more modestly, by this concept, I mean the relational patterns through which resources are allocated within a certain collectivity, decisions taken, collective representations elaborated, feelings of solidarity and mutual obligation forged. In my attempt to build a typology of coordination patterns I focus in particular on two key analytical dimensions, namely, *resource coordination* and *boundary definition*. How can we specify those two dimensions from a relational perspective? By resource coordination I mean the whole set of procedures through which decisions are taken regarding the use of organizational resources – e.g. in relation to the choice of certain campaigns, of specific forms of action, or of certain partners rather than others. I also mean the processes of leadership selection and the definition of criteria for membership. Within specific organizations, such decisions may be taken and implemented through formal as well as informal procedures, although in most cases through a combination of the two (March and Simon, 1958; for a summary, Camerer and Knez, 1996). We can extend this logic, however, to organizational fields, by noting that resource of coordination may also take place through exchanges between informal groups and formal organizations that maintain formal independence and autonomy. In those cases, the focus is on the structure of interorganizational networks, starting from an evaluation of their density (Mizruchi and Galaskiewicz, 1993; Diani, 1995; Diani and Bison, 2004; Ansell, 2003). In some cases, organizations may concentrate most of their resources on their own project and devote a very limited amount of resources to collaborative projects, which results in fairly sparse interorganizational networks. In other cases, resources invested in collaboration may be substantial and are more likely to lead to fairly dense interorganizational networks.

At the same time, processes of boundary definition have also been recognized as central in organizational dynamics (Lawrence and Lorsch, 1967), given their implications for the internal functioning of an organization and the potential for internal conflict and allegiances that they generate (see Camerer and Knez, 1996). Again, what is important within organizations is even more relevant in broader organizational fields. Specifying processes of boundary definition in relational terms is more complex than in the case of resource coordination, as they might be primarily associated with ideational elements, social representations, framing processes. While all of these are crucial, my focus here is on the relational mechanisms that support the creation and reproduction of boundaries. Boundaries are criteria that classify elements of social life in different groups and categories, while shaping the relations between those elements both within and between those groups (Abbott, 1995; Tilly, 2005). They may be seen as the result of processes establishing connections between otherwise separate phenomena, events, or actors (Abbott, 1995, p. 870). As such they can be treated as a result of processes of identity building, which also consist of establishing connections across time and space (e.g. between phases in individual lives, or between different generations, or between events occurring simultaneously in different locations, etc. – Somers, 1994; Melucci, 1996, Chapter 4; Pizzorno, 2008).

Processes of boundary definition and identity building are embedded in relational structures. As Simmel (1964) famously put it, identities are defined by the intersection of group memberships and affiliations. In the context of organizational fields promoting collective action, this means pointing at the role that "latent" networks (Melucci, 1996), connecting different organizations through activists' multiple memberships and personal connections, play in generating broader collective identities and specific boundaries. In all these cases, boundary definition takes place through intense multiple membership and personal connections cutting across organizational boundaries. Joint membership is probably an adequate proxy for primary ties as it is diffuse rather than specific, and expressive rather than instrumental: people join multiple organizations (in voluntary associations) to express their support to them all and to participate generically in their activities, rather than for very specific, ad hoc purposes. And in any case, this is consistent with current references to epistemic communities or communities of practice, where the basis is clearly not kin or friendship. Moreover, others also look at organizational memberships as one of the bases for the ties that make up the fabric of a collectivity (Breiger, 1974; Degenne and Forsé, 1999, pp. 41–44).

A few clarifications are in order before moving to our typology. Firstly, I refer to "modes of organization" rather than to "governance," because I think that the latter should be reserved for forms of coordination that relate to the production of binding decisions, namely to the sphere of the political strictu sensu (Sorensen and Torfing, 2007; Piattoni, 2010). This, irrespective of the fact that in practical terms there are several analogies between some of the network models that I am going to introduce, and the principles guiding network governance. Likewise, there are clear affinities between the model presented below and typologies of economic forms (like Ouchi, 1980- "markets, bureaucracies, and clans" and Podolny and Page's, 1998- "markets, hierarchies, and networks") or broader typologies of social regulation (Streeck and Schmitter, 1985). However, here the focus is restricted to the production of collective action, that can be reframed as those services, broadly defined, that enable citizens to take part in collective action. These refer to both collective action that is expression of a conflict, and to action which is instead driven by goals and motivations around which there is a wide consensus, as solidaristic voluntary action.

## 6.2  Coordinating collective action: social movements, coalitions, organizations, subcultures/communities

Let us start our exploration of modes of coordination by looking at the distinctive traits of social movements. I propose in particular that they be defined by the intersection of *dense networks of informal inter-organizational exchanges* and processes of *boundary definition* that operate at the level of broad collectivities rather than specific groups/organizations, through dense interpersonal networks and multiple affiliations. As far as resource coordination is concerned, several elements are worth stressing. Firstly, social movements consist of a multiplicity of formally independent actors, that articulate the broad goals of the movement as a whole in relatively different issue priorities and specific agendas. These actors may display a huge variety of organizational forms, ranging from the highly formal to the totally loose and spontaneous. They may also differ substantially in the amount and types of resources and skills they command.

At the same time, virtually all of the most significant actions conducted by movements take place through coalition work. Very few organizations actually command enough resources to conduct mobilizations on their own. Moreover, actions promoted by broad coalitions are more likely to attract public attention, be perceived as worthy, and gain political legitimacy. However, this does not mean necessarily that huge encompassing coalitions easily develop. On the contrary, movement actors are normally engaged in highly differentiated joint initiatives and campaigns. These may range from one-off protest events to sustained campaigns spreading over months or possibly years, and may vary substantially in the types of action repertoires adopted, and in the types of social constituencies involved. The format and composition of coalitions is variable, even though some organizations may cooperate on a recurrent basis on different campaigns. Although not all organizations that identify with a given movement collaborate simultaneously on every single project, taken together these coalitions result in relatively dense – at times very dense – networks of interorganizational collaboration.

The terms of interorganizational collaboration are informal, and need to be renegotiated each time a new issue/opportunity/threat emerges. In other words, each collective action event is the product of a specific negotiation. These negotiations refer to several aspects of mobilization campaigns, including the articulation of the specific goals, the choice of the most appropriate tactics and mobilizing messages, the identification of the social sectors from which to search for support, the contributions that each coalition partner is expected to give. This does not mean that practices of repeated collaboration between different organizations may not generate informal routines that reduce the costs of the negotiation, but these routines are not formalized. This resembles closely a definition of network organizations as *"any collection of actors (N > 2) that pursue repeated, enduring exchange relations with one another and, at the same time, lack a legit-imate organizational authority to arbitrate and resolve disputes that may arise during the exchange.* In a pure market, relations are not enduring, but episodic, formed only for the purpose of a well-specified transfer of goods and resources and ending after the transfer. In hierarchies, relations may endure for longer than a brief episode, but a clearly recognized, legitimate authority exists to resolve disputes that arise among actors." (Podolny and Page, 1998, p. 59.)

As far as *boundary definition* processes are concerned, social movements have no formal boundaries and no formally defined criteria for inclusion or exclusion. There are no "social movement members," although members of formal organi-zations often participate in a movement. The only criterion for "membership" is by being part of activities and/or organizations that are associated with the move-ment. By "associated" I mean that they are socially constructed as being linked to a broader collective experience called a "movement." Therefore, the boundaries of a movement are defined by processes of mutual recognition whereby social actors recognize different elements as part of the same collective experience and identify some criteria that differentiate them from the rest. Those elements may be individuals or organizations, but also events.

To begin with, individuals may be associated with a movement to the extent that they recognize each other, and are recognized by other actors, as parts of that particular movement. It does not suffice that they adopt certain lifestyles, hold certain values and opinions, or show willingness to engage in certain actions as individuals. It is necessary that they represent themselves and be represented by others as protagonists of a broader collective process. Likewise, organizations do not belong in a movement because of their issue priorities, their strategies and tactics, or their organizational profile, but because they define themselves as part of that movement, and are perceived as such by significant others. For example, there is nothing in an association for the protection of birds that makes it part of the environmental movement; this depends at the very least on the inclusion of bird protection in a broader environmental agenda and on a recognition of the affinity between that particular organization's goals, agendas, and policies and those of the organizations associated with the environmental movement. The same logic can be applied to events: we do not get a movement out of the sum of the events on issues with similar characteristics, but out of processes of meaningful construction that associate the specific events to a specific project. A protest against industrial pollution in a working class neighborhood may be an instance of the environmental movement if certain representations of reality prevail; but it may also be equally plausibly regarded as an example of class struggle and associated with working class movements, or with localistic, not-in-my-backyard types of movements, if other narratives prevail.

Processes of boundary construction and identity building are essential first of all, because they secure the continuity of social movements over time and space. Social movements exist, in other words, because both actors mobilized in them and (more or less sympathetic or interested) observers are capable of locating in a broader picture actors and events that operate in different points in space and time (e.g. environmentalism exists to the extent that people are capable of providing a common interpretation for actions on nuclear energy, industrial pollution, animal protection, occurring in different localities and at different time points). Boundary construction is also essential because it provides participants with the necessary motivations to act. Although participation in collective action may also be driven by instrumental, contingent calculations or by other mechanisms (e.g. threat, emulation, etc.), identity plays an essential role in supporting actors' decisions to undertake long-term, sometimes dangerous, always uncertain projects. This not only applies to the individual level, on which analysts have focused (Pizzorno, 1978; Melucci, 1996; McAdam, 2003), but also at the organizational level: while interorganizational exchanges are subject to constant negotiation, the establishing of new alliances is easier if there are routines and recurrent practices that also reflect, in particular, identities and definitions of boundaries.

Like resource coordination, boundary definition also often takes a multidimensional, complex form. We only rarely have clear-cut identities and boundaries, neatly separating movements from their environment. Rather, we have boundaries that are often permeable, and more or less dense areas of mutual recognition, and possibly chains of recognition (Pizzorno, 2008). It is important to stress the dual nature of boundary definition, at the organizational level and the movement level. The fact that there is a movement-level identity, that is, a boundary encompassing all the actors associated with a movement, does not mean a demise of organizational identity. On the contrary, feelings of belongingness, solidarity, and obligations may be, and often are, addressed to both specific organizations and a movement taken as a whole (Lofland, 1996, p. 11; see also Zald and McCarthy, 1980). This does not rule out the possibility that some individuals identify exclusively with a movement, without developing loyalty and affiliations to any specific organization (think for example of the affinity groups in the global justice movement described by McDonald, 2002). In general, however, there is a (variably positive or negative) tension between organizational and broader identities. Finally, boundary definition is sustained by activists' multiple affiliations and involvements in several experiences through membership, personal connections, participation in activities (Carroll and Ratner, 1996).

This model elaborates and systematizes on earlier views of movements as networks, most notably put forward by anthropologist Luther Gerlach. In a recent (partial) reformulation of his original position (Gerlach, 1971), he proposes to

Table 6.1. *Modes of coordination of collective action.*

|  |  | Boundary definition | |
|---|---|---|---|
|  |  | Field level | Organizational level |
| Interorganizational networks of resource coordination | Dense | *Social movement* | *Coalition* |
|  | Sparse | *Subculture/ community* | *Organization* |

view social movements as segmentary, polycentric, and integrated networks: "Segmentary: Composed of many diverse groups, which grow and die, divide and fuse, proliferate and contract; Polycentric: Having multiple, often temporary, and sometimes competing leaders or centers of influence; Networked: Forming a loose, reticulate, integrated network with multiple linkages through travelers, overlapping membership, joint activities, common reading matter, and shared ideals and opponents." (Gerlach, 2001, pp. 289–290).

Although Gerlach's insights may not have gained the full recognition they deserved, his views of movements as complex systems of interdependence clearly resonate in later studies of movement networks and organizational fields (e.g. Diani, 1995; Ansell, 2003; Diani and Bison, 2004; Armstrong, 2005; Smith, 2005b).

If the coupling of informal resource coordination through networking and boundary definition processes that span and cut across organizational identities represents the peculiarity of social movements in analytic terms, it does not exhaust the modes of coordination that we can find in any specific empirical episode of collective action – or in any concrete organizational field. Different combinations of our two analytical dimensions generate a fourfold typology that is more representative of the broad variety of "real-life" situations of collective action (Table 6.1). Let us start with *coalitional processes*, the most tricky as social movements and coalitions are frequently used interchangeably. In terms of resource coordination, coalitional processes are actually very similar to social movements, for the reasons stated above: they consist of multiple, often heterogeneous, independent actors, sharing resources in pursuit of some shared goals. Social movements are usually seen as strings of coalitions (e.g. Meyer and Corrigall-Brown, 2005; Van Dyke and McCammon, 2010).

And yet, the boundary definition process on which coalitions are founded is temporary and locally circumscribed, which is not the case for social movements. It is still necessary to define a collective we and a collective them, yet those definitions do not span time and space as it happens in the case of social movements. They are mainly driven by circumscribed, instrumental preoccupations, regardless of the more or less moral or strictly economic nature of their goal. Throughout the process of collective action, participants' loyalties and priorities remain firmly within the boundaries of specific organizations, and there are no attempts to forge broader and deeper bonds. With all due adaptations, a similar difference can be found in inter-organizational studies in management between ad hoc alliances and longer-term interorganizational networks (see e.g. Barringer and Harrison, 2000).

Likewise, and regardless of whether their aim is stopping a policy proposal or discouraging some kinds of behavior, or pushing for alternatives, coalitions exhaust their function when their goal is either achieved, or when it is clear that the cause has been lost. There is nothing left over from a coalition in terms of feelings of belongingness to a broader collective entity, or of attempts to build a longer-term and more solid collective identity by linking the specific campaign to larger collective projects, encompassing multiple actors.

This view of coalitions is best expressed by social scientists that are (or were, like Gamson) closest to the rational choice paradigm. They define them as " ... a group of players who are able to make binding agreements to implement agreed strategies" (Hargreaves-Heap *et al.*, 1992, p. 95), or "... temporary, means oriented alliances among individuals or groups which differ in goals...[with] little value consensus...[and] tacit neutrality on matters which go beyond the immediate pre-rogatives" (Gamson, 1961, p. 374). Both definitions stress the short-term objectives of coalitions and make no reference to longer-term goals and broader values. As Kadushin *et al.* (2005, p. 258) put it, "A coalition is defined in economics and political science as joint action among two or more parties to achieve a common goal ..., or alliances that are temporary and fluid, dissolving or changing as goals or members' self-interest is re-defined."

It is certainly true that in social movement, community studies, or advocacy coalition literature, a more inclusive view of coalitions has developed, stressing deeper solidarities and broader, value-laden goals (Kadushin *et al.*, 2005; Van Dyke and McCammon, 2010). And it is similarly true that it is often difficult to tell the difference between a movement and a coalition when observing con-crete cases of mobilization. It is for instance often the case that coalitions evolve gradually into fully-fledged movements, establishing connections to related cam-paigns and broadening their perspectives. However, the difficulties in the empirical operationalization of concepts (on this see e.g. Diani and Bison, 2004) should not lead us to blur analytical differences. Doing that entails serious interpretative problems when looking at concrete mobilizations. Let us think for instance of the demonstrations that challenged the imminent war against Iraq in February 2003 (Walgrave and Rucht, 2010): were they primarily the expression of peace movements, i.e. the peak of long-term sustained mobilizations promoted and/or led by organizations whose primary identities and solidarities lay with the peace, anti-war, anti-military cause over a long time span? Or were they, alternatively, mainly the expression of ad hoc coalitions, bringing together people and organizations whose primary concerns focused on other agendas, and whose goals did not go beyond expressing people's worries at a particularly unpopular decision by their governments, or who perhaps regarded those demonstrations as part of broader anti-governmental strategies? It is clearly not the same thing. This does not mean that in each particular episode of collective action one will not find more than one process at play; but it is important to recognize their distinctiveness in order to better assess their interplay, rather than collapsing them under broader, connotationally poorer, categories (Sartori, 1970).

Another important point to keep in mind is that a huge chunk of protest activity, as well as voluntary action (actually the largest part of it), follows the logic of *organizational processes*. These consist of modes of resource coordination and boundary definition that ignore inter-organizational networking and largely take place within specific groups or organizations. This model differs from otherwise cognate models developed in the context of typologies of social regulations and governance such as "hierarchy" (Podolny and Page, 1998) or "bureaucracies" (Ouchi, 1980), in that it does not reflect necessarily the traits of the Weberian bureaucracy. It actually accommodates organizational forms that range from the extremely hierarchical and formalized (such as twentieth century mass parties) to the extremely decentralized and informal, such as alternative communes or grassroots groups; from the extremely endowed with resources, such as business associations, to the extremely deprived, such as neighborhood action groups. What matters here is that action is promoted and coordinated by units that have an autonomous decisional capacity, whether this comes from formally appointed leaders or officers, or from the grassroots, the participatory deliberations of activists' meetings.

This model is not necessarily a better fit to organizations with a weak connection to protest politics. With the exception of revolutionary parties of the left and right, political parties, for example, or business associations, may engage only sparingly in protest and other participatory actions, however, when they do – e.g. in order to challenge a hostile government – the amount of resources they command may enable them to go on their own. However, while the amount of available resources may be one important consideration behind decisions not to engage in sustained alliance building, other factors may also matter. Regardless of their size and strength, organizations may think of themselves as in competition with their like for scarce resources. They may therefore subordinate alliance building to their organizational needs, such as the strengthening of the peculiarity of their profile (alliance building inevitably implies compromises and may render it more opaque), or that of securing a specific niche by becoming quasi-monopolist on specific issues. Many organizations working on public and collective interests or engaged in advocacy on behalf of other groups operate precisely on the basis of this logic. In some cases this leads to more or less negotiated forms of division of labor (in the case of environmentalism, e.g. with organizations specializing in single issues such as transport, energy, etc., or animal rights groups focusing on different species). In other cases organizations compete explicitly for the same constituency. When ideological resources are heavily used in such competition in order to increase the comparative worth of the different competing organizations, this may encourage factionalism and sectarianism (see e.g. Della Porta, 1995 on left-wing terrorism).

The reference to sects draws our attention to the second analytical dimension, boundary definition. They remind us in particular that the definition of boundaries and identity building is by no means less relevant in instances of collective action coordinated through an organizational model, than in the case of social movements. The difference lies in the fact that loyalties and attachments are largely, if not exclusively, focused on specific groups and organizations rather than on broader collectivities. The exclusiveness and rigidity of these boundaries may vary substantially: boundaries may be very loose, as in the case of large environmental associations such as WWF or Greenpeace, where membership is prevalently of the chequebook type and adhesion implies no subscription to fundamental values; or they may be very tight, as in the case of neo-religious sects like the Hare Krishna, fundamentalist political organizations, radical political organizations of the left and right. Membership in one group may entail the possibility of adopting multiple identities, reflective of one's multiple interests and commitments, or may require full identification to the detriment of any other loyalty. The reference to sects draws our attention to boundary definition. Definition to boundaries and building of identity are key factors in collective actions coordinated through an organizational model, and in the case of social movements. However, groups of organizations differ from broader collectivities because of stronger levels of loyalty and attachment.

Finally, it is also possible to think of *subcultures/communities* as a distinct mode of coordination of collective action (see also Ouchi, 1980's related concept of the clan). By this I mean a process in which interorganizational linkages are sparse, yet there are widespread feelings of identification with a much broader collectivity than the one represented by specific organizations, and a set of practices, multiple affiliations, etc. that support this. In the sociological tradition, "community" was characterized by (a) a set of actors sharing a distinctive sense of belongingness and norms, (b) being located in a territorially defined area

(Wellman, 1979; Knoke, 1990). The emphasis on values was also strong among proponents of the concept of subculture, as a subgroup within society defined by specific lifestyles (Yinger, 1960, p. 626). Critics from a relational perspective have contended that emphasis should be placed instead on the distinctive relational patterns that link actors to each other, based on primary ties (kinship and friendship in the first instance), and that reproduce themselves even when scattered over larger territories (Wellman, 1979), and even in cyberspace (Wellman and Gulia, 1999). It has also been suggested that the notion of community does not even necessarily imply diffuse rather than specific relations: epistemic communities, or communities of practice, are characterized by a distinctive set of interests and/or moral orientations around which interaction develops. They are not functionally designed in relation to specific goals, like bureaucracies, but rather as an expression of particular orientations, interests, lifestyles.

The reason for the lack of dense interorganizational networks may be diverse. In some cases, they may originate from dynamics of interorganizational competition, factionalism, and sectarianism analogous to those described above, yet taking place in settings which are characterized by shared cultures and significant boundary definition. For example, in spite of deep interorganizational conflicts between new and old left political organizations, and even among new left ones, Italian collective action in the 1970s was still embedded in diffuse networks of personal acquaintance and exchange between activists, personal participation in activities promoted by other organizations, and a broader definition of "the left" that somehow kept together its different and mutually contentious components.

In other cases, interorganizational exchanges may be sparse because organizations focusing on collective action are simply non-existent, or alliance building is discouraged by the high costs associated with a repressive regime. For example, accounts of collective action in Islamic countries of North Africa and the Middle East fairly consistently point at the role of the mosques and the bazaar in providing the community settings in which collective action is organized, in the absence of specifically built organizations (Bennani-Chraïbi and Fillieule, 2003). Similar indications emerge from accounts of the structures supporting political dissent in the socialist regimens of Eastern Europe before 1989 (Johnston and Snow, 1998; Glenn, 1999). At other times, interorganizational coalitions are missing because that type of campaigning is not consistent with the characteristics of existing organizations. For example, within critical subcultures that express themselves mainly through the practice of specific lifestyles, such as gay and lesbian communities, many organizations have cultural rather than political mobilization goals – even allowing for the difficulty of drawing a clear line. In those cases, boundary definition in the community takes place through the multiple involvement of people in cultural associations and events, in community forms of life, etc. They also provide from time to time the setting to coordinate political mobilization, but do not operate to promote directly extended interorganizational alliances. The networks of individuals focusing on cultural practices and specific lifestyles, like the feminist and environmental activists studied by Melucci (1996) in Milan, are certainly a case in point. So are the "affinity groups" bringing together radical opponents of neoliberal globalization (McDonald, 2002). Yet another type of subcultural mode of coordination may be found in transnational initiatives such as the so-called 'dynamic coalitions," comprising individuals interested in Internet governance issues and active largely on the basis of their personal skills and orientations, rather than as representatives of specific organizations (Pavan, 2012).

## 6.3 An illustration: Milanese environmentalism in the 1980s

Network data collected among 42 environmental organizations in Milan (Diani, 1995) provide an example of how different relational patterns of collective action may be found within a specific organizational field that both external observers and internal participants broadly, if vaguely, perceive as a "movement." Here I look at the distribution of ties within and across different clusters of organizations. These clusters are defined on the basis of a nominalist principle (Diani, 2002, p. 176), more specifically, by a combination of function, issue, and location,[2] and they correspond (Figure 6.1) with

- the core environmental associations (leftist Legambiente and the more moderate WWF and Italia Nostra);
- animal rights organizations;
- conservation and nature protection associations;
- the Milan Green List and other political ecology organizations;
- local groups acting respectively in the western, southern, and northern Milanese periphery.

Of these clusters, only the core groups, animal rights, and West and South Milan local groups displayed significant levels of resource coordination among themselves in relation to the promotion of joint campaigns. These are reflected in the ratios between the observed ties and the ties one would expect in the case of random distribution of links (Table 6.2).[3] Levels of coordination were similarly sparse among the other political ecology and conservation groups. The overall structure of

---

[2] In the original study positions within the network were identified with a realist (structural equivalence) procedure. I use a nominalist approach here as results are easier to interpret. However, the profile of movement networks that emerges from the two analysis does not differ significantly (see Diani, 1995, p. 123 ff.).

[3] I consider as a sign of relatively high density of ties those cases in which the ratio observed/expected exceeds one standard deviation above the mean value.

Table 6.2. *Ratio of observed/expected resource exchanges[a].*

|   |                    | 1     | 2    | 3    | 4    | 5    | 6    | 7    |
|---|--------------------|-------|------|------|------|------|------|------|
| 1 | Core organizations | 14.84 | 0.82 | 0.00 | 3.96 | 0.00 | 1.98 | 0.00 |
| 2 | Animal rights      | 4.95  | 6.43 | 2.06 | 0.49 | 0.00 | 0.99 | 0.21 |
| 3 | Conservation       | 0.82  | 1.65 | 0.99 | 0.99 | 0.00 | 0.49 | 0.00 |
| 4 | Political ecology  | 8.91  | 0.00 | 0.00 | 0.74 | 0.59 | 1.19 | 0.25 |
| 5 | West Milan         | 2.97  | 0.00 | 0.00 | 0.59 | 2.97 | 1.19 | 0.25 |
| 6 | South Milan        | 4.95  | 0.99 | 0.49 | 1.78 | 0.00 | 5.94 | 0.00 |
| 7 | North Milan        | 1.65  | 0.00 | 0.00 | 0.25 | 0.25 | 0.00 | 1.69 |

[a] Overall network density: 0.067; mean = 1.09, SD = 1.75

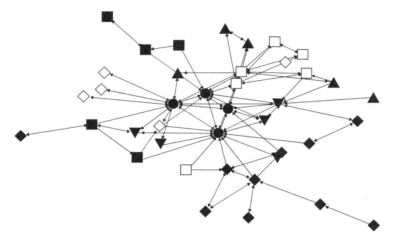

Figure 6.1. Interorganizational exchanges between environmental organizations (Milan, 1980s). Key to symbols: ● = Core environmental associations; □ = Animal rights groups; ■ = Local groups, West Milan; ▲ = Conservation groups; ▼ = Political ecology groups; ◊ = Local groups, South Milan; ◆ = Local groups, North Milan.

resource exchanges turned out to be highly centralized, with the three main environmental associations strongly connected to each other and also strongly connected to most of the other clusters, exceptions being conservation groups and local groups in the northern periphery (Figure 6.2; see also Diani, 1995, pp. 121–126).

Boundary definition, reflected in overlapping memberships and personal ties between core activists, further qualifies but does not drastically alter the previous picture. The overall structure is similar to the one detected for resource exchanges, the only difference being that conservation organizations are connected to the core

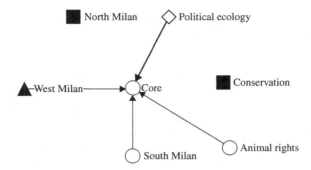

Figure 6.2. Resource exchange ties, pre-defined blocks (environmental groups in Milan, 1980s). Key to symbols representing different modes of coordination within blocks: ○ = social movement; ▲ = coalition; ■ = organization; ◊ = subculture.

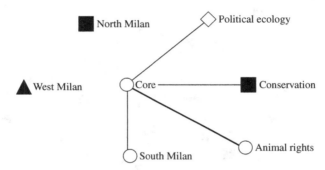

Figure 6.3. Boundary defining ties, pre-defined blocks (environmental groups in Milan, 1980s). Key to symbols representing different modes of coordination within blocks: ○ = social movement; ▲ = coalition; ■ = organization; ◊ = subculture.

through personal contacts, while local groups in West Milan are not (Figure 6.3).[4] Moving to the position of specific clusters, three of those that were densely connected in terms of resource exchanges were similarly connected by their core activists' multiple involvements; and, therefore, closest to a "social movement" relational process (Table 6.3; see also Diani, 1995, p. 133). In contrast, local groups in the Western periphery showed no "boundary-defining" linkages between them, suggesting that in their localities they were operating mainly according to a "coalitional" logic of action.

Of the three structural positions that showed little internal density of resource coordination, political ecology groups displayed relatively high levels of joint memberships and interpersonal contacts. This suggests that they were closest to

---

[4] The use of a stricter criterion to assess density of ties accounts for the differences between this and earlier analyses of the same data (Diani, 1995, p. 133).

Table 6.3. *Ratio of observed/expected boundary defining ties[a].*

|   |                     | 1     | 2    | 3    | 4    | 5    | 6    | 7    |
|---|---------------------|-------|------|------|------|------|------|------|
| 1 | Core organizations  | 12.34 | 4.11 | 2.57 | 3.09 | 0.62 | 3.09 | 0.77 |
| 2 | Animal rights       | 4.11  | 6.17 | 1.54 | 1.54 | 0.31 | 1.23 | 0.00 |
| 3 | Conservation        | 2.57  | 1.54 | 0.00 | 0.31 | 0.00 | 0.62 | 0.00 |
| 4 | Political ecology   | 3.09  | 1.54 | 0.31 | 3.70 | 0.74 | 1.48 | 0.31 |
| 5 | West Milan          | 0.62  | 0.31 | 0.00 | 0.74 | 0.00 | 0.37 | 0.00 |
| 6 | South Milan         | 3.09  | 1.23 | 0.62 | 1.48 | 0.37 | 7.41 | 0.00 |
| 7 | North Milan         | 0.77  | 0.00 | 0.00 | 0.31 | 0.00 | 0.00 | 1.68 |

[a] Overall network density: 0.086; mean = 1.08, SD = 1.17

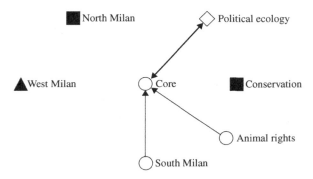

Figure 6.4. Multiplex ties, pre-defined blocks (environmental groups in Milan, 1980s). Key to symbols representing different modes of coordination within blocks: ○ = social movement; ▲ = coalition; ■ = organization; ◇ = subculture.

reflecting, as a distinct subfield, a subcultural/community mode of coordination: they might not have been working closely with each other but were nonetheless embedded in deeper ties created by their active members. Most organizations operating in peripheral areas showed no particular density of ties generated by their activists' multiple involvements, to suggest a conception of collective action mostly focused on single projects. While organizations in West Milan at least showed some dense resource exchanges, which reflected a coalitional mode of coordination, in North Milan this was not the case as collective action seemed to be mainly coordinated within specific organizations. The same applied to conservation groups, that seemed to represent an aggregate of independent organizations with an interest in environmental issues, rather than a distinctive sector of a broader movement.

We gain further insight if we look at the multiplex ties generated by the overlap of resource exchanges and boundary defining ties (Figure 6.4). These ties are the closest approximation to a social movement mode of coordination. Their distribution within the field identifies a star-shaped component plus three clusters without significant multiple connections to the rest of the field. The structure of the main component suggests a fairly centralized social movement, in which the core organizations play the key role, providing the link between animal rights and political ecology groups, most notably the Green party.

At the same time, there is quite a degree of segmentation, involving particular organizations (mostly, informal or semi-formal groups) focusing on specific areas of the city and its immediate hinterland. One set of organizations operates in an area (North Milan) with a strong tradition of left-wing militancy. The persistence of the left-right cleavage in the area contributes to the fragmentation of the ties between this set of organizations. It makes the distinction between political ecology and conservation groups deeper than it would be in settings with a more pacified political culture. Conservation groups also turn out to be largely isolated from the rest of the field in terms of multiplex ties. The other organizations missing multiplex ties to the rest of the field are located in a largely affluent, bourgeois area (West Milan), with little tradition of grassroots radicalism.

It is also worth noting that modes of coordination *across* different clusters of organizations do not fully match those operating *within* clusters. In particular, one of the blocks connected by multiplex ties to the core environmental organizations ("political ecology") is only connected internally by joint memberships and interpersonal friendships. There seem to be distinctive identities and common boundaries between organizations with a distinctive approach to the environment, that however do not translate into dense interorganizational collaborations.

The overall picture suggests the opportunity of decoupling analytically traits of activisms that are usually strongly associated. It seems in particular difficult to sustain the conventional view of the bureaucratic, professionalized, sector of big associations linked by largely instrumental ties into ad hoc coalitions, opposed to the participatory sector of loosely structured, grassroots informal groups, that embed their collaborations in deeper, stronger interpersonal ties. This may perhaps work for some instances of collective action such as the GJM (McDonald, 2002; Smith, 2005a), although even that assumption might be worth further scrutiny. But it cannot definitely be assumed as a general mode of functioning for organizational fields consisting largely of grassroots organizations.

Indeed, while the environmental organizational field in Milan reflects different relational logics, that characterize the functioning of its internal components, it can also be seen as an integrated system of exchange. The integration is due to the role of a restricted number of core organizations through which not only the largest portion of resource exchanges, but also of boundary defining, take place. In other words, the identity of the movement sticks together (no matter how weakly: Diani, 1995), thanks to the fact that a small number of core organizations are regarded as reliable and trustable partners by the large majority of environmental groups that are not always linked directly.

The absence of multiplex ties connecting different sets of organizations without going through the core ones points to a star-shaped structure that is fairly unstable and would be subject to disbanding in the case of internal core deterioration. This departs substantially from the adaptability and resilience of Gerlach's (1971, 2001) policephalous SPIN model, and urges recognition of the plurality of forms that one may find within organizational fields conventionally associated with the category "social movement."

## 6.4 Modes of coordination, network organizations, and policy processes

The typology of modes of coordination, introduced in the previous pages, has several analytic implications. Here I focus on two, namely, its relation to the concept of network organization, elaborated by organization theorists (Podolny and Page, 1998; Kenis and Knoke, 2002; Brass *et al.*, 2004; Provan *et al.*, 2007; Provan and Kenis, 2007), and its consequences for our understanding of policy processes. On the first ground, there are undoubtedly analogies, as both "coalitional" and "social movement" modes of coordination are characterized, similarly to network organizations, by patterned forms of interaction between formally independent units (Jones *et al.*, 1997, p. 914; Monge and Contractor, 2003, pp. 220–221). Likewise, in business networks as well as in social movements, social control passes largely through relational mechanisms such as mutuality, reciprocity, etc. rather than contract: "[implicit and open-ended] contracts are socially – not legally – binding" (Jones *et al.*, 1997, p. 914; see also Ouchi, 1980; Powell, 1990, pp. 304–305).

At the same time, one also has to recognize some important differences. The main one revolves around the difference between a network mode of organization and a "network organization." In the business and management literature there is considerable variation in the focus of analysis, that stretches from networks of formally independent organizations within a broader organizational field (e.g. in Kenis and Knoke's [2002] concept of field-net) to clearly bounded organizations in which the network component refers to the pattern of allocating and coordinating resources between different units (e.g. Baker, 1992). It is the use of the concept in relation to the latter that has prompted some analysts to question the analytic value of the concept of "network organization" (Borgatti and Foster, 2003).

One apparent paradox is that in the business world it is equally possible to find instances of "boundary-less" organizations, in which "where one organization begins and the other ends is no longer clear" (Monge and Contractor, 2003, p. 220), and instances like the classic IBM and Benetton described by Castells (1996), in which boundaries are quite clearly defined: possibly not as clearly as in a hierarchy/bureaucracy, but surely far more clearly than in a loose movement network.

It is also worth mentioning that, by comparison to movement networks, network organizations seem to involve a higher degree of deliberate planning and design (Baker, 1992; Provan *et al.*, 2007), although this may vary across situations (being higher for instance in the case of Benetton than in the case of the networks of collaboration between small firms in Italian industrial districts). The emergence of a network organization does not develop in linear form, e.g. the need to react to specific needs may start the alliance building process, but the final outcome in terms of the emergence of a network model is rarely anticipated (Powell, 1990, pp. 322–323). Secondly, while movement integration relies almost entirely on mutual social control and shared understandings, which results in very loose coupling (Jackson, 2006), business networks are often kept together also by contracts (Provan *et al.*, 2007, p. 482).

To limit terminological confusion, it is important to stress that *within* movement networks one can frequently identify proper "network organizations" that operate as autonomous decision making units. Examples are numerous, including global organizations like Attac (Kolb, 2005) or CRIS – Communication Rights in the Information Society campaign (Thomas, 2006), networks originated from national or local independent groups, and associations to coordinate specific campaigns (e.g. the Globalised resistance campaign in the UK), or umbrella organizations such as the Councils for voluntary sector set up in British cities to provide services to the voluntary sector. Analytically, however, the network organization seems to me a spurious intermediate form between a social movement and an organizational mode of coordination, rather than a distinctive analytic category.

Moving to the implications of this typology for our understanding of policy processes, they may be of at least two different kinds. Recognizing the multiplicity of modes of coordination within organizational fields may affect our perception of different actors as legitimate representatives of a broader field; and it may also affect our expectations regarding the impact of collaborations between authority and the voluntary sector in the implementation of public policy. Taking the Milanese environmental field as an illustration, the reconstruction of its structure confirms in the first place the strength of the claim made by the major national environmental organizations to be actually representative of Italian environmentalism taken as whole. There are no significant flows of resources running through the environmental field that are not channeled through the major core organizations; even processes of boundary definition, reflected in multiplex ties, have the major organizations at their core. Given such configuration of ties, it is obvious that no policy network is conceivable that exclude the major organizations or assign them a marginal role.

At the same time, several elements of caution also emerge from the analysis. Firstly, there are many organizations within the field that have no significant connections to the core. Conservation and nature protection groups are largely disconnected. Most importantly, they tend not to cooperate with each other, but rather to act independently. The presence of several actors trying to strengthen their control over a specific niche of sub-issues means that the policy process should aim to include as many meaningful actors as possible, rather than assuming that major organizations should be oligopolistic representatives of the entire sector – they clearly are not.

It is also worth stressing the presence of many grassroots organizations, located in West and North Milan, among those that operate without strong ties to the core environmental organizations. Once again, in a situation displaying these structural patterns, any attempt to include environmental organizations in the policy process should not be limited to the major organizations. At least judging by the overall flows of ties within the field, core organizations do not seem representative of the sector as a whole to the point of ignoring local action networks, lest the risk of serious difficulties in both the policy formation and – most important – the policy implementation process.

Finally, we should also note that the strong centralization of the overall network is a potential source of risk for the integration of the overall field. It is certainly true that communication is easy and unexpensive thanks to the linking role of the core groups, and that the ideological heterogeneity of the core facilitates the convergence, in the same star-shaped structure. However, this feature also exposes the field to serious risks of disintegration, as there would be limited lines of communication with risks for the solidarity within the core cluster to fall apart. In terms of the policy process, this might result in the short-term in greater opportunities for organizations specializing on specific issues to gain access to policy-makers. But it would be very difficult to identify actors capable of speaking on behalf of environmentalism as a whole, and thus giving the issue a more authoritative representation within broader policy networks.

# 7

# Why skewed distributions of pay for executives is the cause of much grief: puzzles and few answers (so far)[1]

BRUCE KOGUT AND JAE-SUK YANG

One of the most important questions for public policy that arose from the financial crisis that began in the United States in 2007 is the effect of executive pay on risk taking. The regulatory implications of this claim have been significant. Federal Reserve Chairman Ben Bernanke described the Fed's efforts to develop rules that will "ask or tell banks to structure their compensation, not just at the top but down much further, in a way that is consistent with safety and soundness – which means that payments, bonuses and so on should be tied to performance and should not induce excessive risk" (Wall Street Journal, 5/13/2009). The Dodd–Frank Wall Street and Consumer Finance Act, passed in July 2010, included several provisions to give shareholders a say over pay, and to strengthen the risk and compensation practices of boards of directors. Sheila Bair, chairman of the FDIC, said: "This proposed rule will help address a key safety and soundness issue which contributed to the recent financial crisis: that poorly designed compensation structures can misalign incentives and induce excessive risk-taking within financial organizations."[2]

The justification for imposing risk on executives (and workers in general) derives from the wide acceptance of the principal–agent model that informs law and economics and managerial practice. The standard principal–agent model recognizes that the effort of the executive (the agent) is not observable by owners (the principal), and thus compensation contracts use incentive schemes to motivate executives who otherwise would exert less than the contracted effort. For the kind of work that

---

[1] We would like to thank Professors Sudhakar Balachandran and Jerry Kim for permission to use liberally our joint work on executive pay and to acknowledge the financial support of the Sanford C. Bernstein & Co. Center for Leadership and Ethics at Columbia University. Also, we owe a special thanks to Balázs Vedres for his encouragement and patience.

[2] Reported by Reuters, February 8, 2011; http://www.complinet.com/dodd-frank/news/articles/article/fdic-proposes-to-defer-bonuses-to-prevent-excessive-risk-taking.html

---

Networks in Social Policy Problems, eds. Balázs Vedres and Marco Scotti. Published by Cambridge University Press © Cambridge University Press 2012.

executives do, the measurement of effort is difficult to ascertain and incentives are invariably attached to signals, such as stock prices, that are noisy or imperfectly correlated with "true" performance. The problem remains then of figuring out how much incentive pay needs to be promised to get someone to do something. Since the principal will never know the optimal contract under imperfect information, the second-best contract involves shifting a degree of the risk (e.g. a share of profits) onto the agent. It is conventionally assumed that the principal can more efficiently diversify this risk than the agent. In fact, the agent is prevented from diversifying the risk by prohibitions in the contract in order to motivate the agent to avoid "hidden actions" and to act in the interest of the principal.

The classic question in the law and economics literature is how to motivate the agent through the assumption of risk. The public policy question has become the inverse: how to prevent too much risk taking by executives, especially in industries where there are severe externalities. Nuclear energy plants are a good stylized example. Clearly, we want executives to be incentivized to make the most efficient decisions in the production of nuclear energy, but we do not want them to be overly incentivized given the *externalities* of nuclear fallout. Similarly, given that financial bankruptcy can often be contagious (i.e. there is systemic risk), the public policy perspective is to balance risk-taking incentive contracts against the threat of potential negative externalities through engendering a financial crisis.

Whereas the above policy question has focused on the *structure* of pay and the incentives for taking on too much risk taking, the popular debate in the United States and elsewhere has been fueled by the large disparities in the *levels* of income, especially during a time when those who appear to be responsible for the crisis have fared better than the general public. The United States has experienced, since the 1980s, a growing inequality of income due to stagnating wages for the middle-class worker with less than a college degree, and the rising share of income earned by the top 1% of the US workforce. This top percent includes athletes, movie stars, entrepreneurs, as well as professionals, namely executives and lawyers (Kaplan, 2008). Gordon and Dew-Becker (2008) estimate that top executives of the largest 1,500 public firms earned 22% of income in the top 0.01% bracket; this percentage does not include hedge funds and private equity managers whose income grew much faster than that of public company managers, as noted by Kaplan and Rauh (2010). Unlike athletes and stars whose high incomes often decline during life, executives are often paid high incomes as salary and as a share of corporate value, such as through stocks and stock options, which while variable, are more enduring sources of wealth.

The explanation for why executives are paid so much has moved from justification by incentives to that of talent. Nor is this language without significance, for by being classified as "talent," the salient comparisons for top managers are those individuals who are talented in the more classic sense, such as athletes and stars.

The chief executive officer (CEO) of a publicly traded company is responsible for managing large and complex organizations. These responsibilities include setting the strategy of the firm, administering and leading a hierarchy of many employees, raising financing from capital markets, and deciding whether to expand by growth or acquisition, or be acquired. These tasks are difficult and require skills and the competence of well-educated and experienced individuals whose numbers are not abundant. Consequently, one of the principal functions of a board of directors is to reward CEOs sufficiently to attract and retain them. In the parlance of economics, these rewards must satisfy their "reservation" salary, which is the value an executive would require to accept the position rather than accept work elsewhere.

All economic theories of executive compensation accept this basic logic, and so stated, the logic is consistent with many sociological theories. However, there are large disagreements over the question whether firms pay their executives more than this reservation wage. The argument that executives are paid their reservation wage has been elegantly advanced by Gabaix and Landier (2008). They argue the following. If labor markets for executives were competitive, markets would clear by "assortative matching" through which more talented CEOs would work for firms that, because of the specific complexity of the required job, are willing to pay a premium for these talents; less talented CEOs would work for firms that are less demanding. In equilibrium, all CEOs are paid their reservation salary and ranked in order by their talent.

The evidence for this "best athlete" argument is the simple correlation of two power law distributions: the log market value of firms and the log salary of CEOs, a correlation which is easily apparent in the plot of the data. In Kim *et al.* (2011), we have plotted these data for the time period of 1996 to 2009 for the chief executive officers of American public corporations, using the Compustat and Execucomp databases. While the regression fit is less than 0.5 $R^2$, there is nevertheless a Gaussian-shaped cloud around the least squares line. If the assumption that larger firms require more talented CEOs, this plotted relationship supports a theory that better athletes work for bigger firms and are paid more than those less capable CEOs who work for smaller firms. For critiques of high pay of finance firms, this simple plot shows that CEOs of finance firms are paid well because their firms are big. However, there is more to this than meets the eye, a point to which we return below.

The appeal of this theory, that big firms hire the better talent, is its consistency with an important stylized fact, namely that since the 1970s average CEO pay has risen dramatically and the size of firms has also increased. The correlation of pay and size makes sense of this fact. Since firm size has increased, executive pay has as well. If we take this evidence at face value, the reported regression still accounts for less than 50%, which is interesting since the correlation is between two logged economic series that can be expected to be positive simply due to common trends. Moreover, as noted by Gordon and Dew-Becker (2008, p. 23), executive pay has in fact risen almost three times faster than firm growth in the past decades. While firm size (market value) and pay are correlated, it is statistically an open question as to why the latter has risen so much faster.[3]

In the following pages, we review, and analyze, the data on the structure and level of pay in regard to five aspects of executive pay:

(1) What is the recent evidence that the structure of compensation (i.e. the use of stocks and stock options) mattered to risk taking by financial firms?
(2) How has the structure of pay differed by industry?
(3) What is the evidence that topological properties besides industry classification matter for the level of pay through a diffusion process?
(4) Given the sparseness and fragmentation of board interlocks, has mobility (i.e. the movement of CEOs or executives) mattered towards diffusing high pay?
(5) Finally, does the gender of the CEO (and of the directors to a board) matter to understanding pay levels, that is, is there evidence for discrimination that influences the pattern of pay?

We turn to each of these questions below.

## 7.1 Why incentive pay is a poor explanation for high levels of executive income

Starting in the 1980s, the structure and composition of executive pay in the United States underwent a revolution, moving away from a high proportion of remuneration in the form of fixed salary to a greater reliance on "pay for performance." The data analyzed in Balachandran *et al.* (2010) shows clearly the change in the composition and levels of pay. Not surprisingly, the period of the late 1990s through 2001 shows a large increase in stock options due to the rapid increase and then collapse of stock market prices in general. This change in the structure of pay means that there was a greater use of incentivized compensation.

The theoretical justification for relying on incentives is given by the principal agent model described above. The principal agent model was conceived to explain the design of contracts when information is imperfect and goals are not a priori aligned. Most of the evidence that the theory has predictive value draws upon studies of productivity and incentives, such as for taxi cab drivers or fruit pickers or experiments in labs, often using students. There are good reasons to think that these models do not work well for highly paid managers.

The explanation for these defects of theory regarding high levels of pay rests on the difficulty of incentivizing already rich managers. The standard model for moral

---

[3] See Gordon and Dew-Becker (2008), on the unrealistic prediction that arises from the Gabaix–Landier model, given the empirical evidence on the elasticity of pay with reference to firm market value.

hazard qua hidden action is to characterize the manager's utility by a constant absolute risk aversion (CARA), i.e. $U(x) = 1 - e^{-ax}$. As is well known, such a model means that $1,000 of extra pay is equivalent for the executive earning a $100,000 and for the executive earning $10 million. In the context of an executive compensation, constant relative risk aversion utility, e.g. $U(x) = log(x)$, is more satisfactory in that executives might care more about proportional increases in income than absolute increases. However, a relative risk aversion specification has very important and non-trivial effects on the level of pay and the required incentives. Imposing utility functions that are separable in effort and income is also not innocuous in the context of large compensation packages. CARA utility functions are needed to find closed-form solutions and simple linear descriptions of pay incentives (Bolton and Dewatripont, 2005).

However, the mathematical specifications carry considerable limitations for understanding executive pay. The problem with these models is an under-appreciation of the massive incentives required to motivate already highly paid and wealthy CEOs. If pay performance compensation is to incentivize managers, then the implication of the large incentive packages for CEOs of financial institutions is that top managers will experience a decline in marginal utility in income. Since an additional dollar has less utility for a CEO than for a fruit picker, it is necessary to incentivize pay to increase income and wealth. This observation provides a different though not inconsistent view to that of the talent theory, that compensation grows in firm size due to the need to attract people to manage complexity. In this way, executive pay grows in order to incentivize managers to work hard, because they are already well paid.

If we turn back to the financial crisis in the USA, it is not surprising that managers seek to eliminate the undiversified risk accrued through holding stocks and stock options in the company that employs them. Leading up to the crisis, top managers had exercised their options and reduced their exposure to the performance of their company. Bebchuk and Spamann (2010) calculate that the top management at Bear Stearns and Lehman cashed out $1.3 billion and $1 billion, respectively, between the years 2000 and 2008.

Still, even if executives cash out, the evidence points to a surprisingly high level of wealth invested by top management in their companies and yet the executives still took on a lot of risk. A study by Fahlenbrach and Stulz (2011) finds that the top 20 CEOs had equity stakes valued at more than $100 million; the mean (median) value of the CEO's equity stake was $88.1 million ($36.3 million). The average equity held by top management was about 1.6% of total shares. This low percentage indicates the challenges of executive compensation of bank managers who are already wealthy. The aggregate compensation paid by public companies to their top five executives during the considered period added up to about $350 billion, and the ratio of this aggregate top-five compensation to the aggregate earnings of these firms increased from 5 percent in 1993–1995 to about 10 percent in 2001–2003 (Bebchuk and Grinstein, 2005). So clearly, top executives were paid a lot of incentivized compensation in the form of stock and exercised options.

Did this pay lead to better performance of their financial firms? Fahlenbrach and Stulz (2011) note that the top five best paid executives in financial services in 2006 were the CEOs of Lehman Brothers, Bear Stearns, Merrill Lynch, Morgan Stanley, and Countrywide Finance. Only one of these companies survived the crisis as an independent operation.

More thorough empirical analyses confirm this anecdotal evidence that highly incentivized managers took on excessive risk. Balachandran *et al.* (2010) assembled panel data on 117 financial firms from 1995 through 2008, using the financial crisis as a type of "stress test" experiment to determine the relation of equity-based incentives to the probability of default. They first estimated the default probabilities using a specification that allowed for time-varying parameters to a Black–Scholes–Merton option model. These estimations provided a time series on the implied probabilities that the market predicted for default.

For example, in Figure 7.1, we show the estimates of the implied risk of default for two banks, JP Morgan Chase and Wells Fargo, and for an investment bank, Goldman Sachs, that converted to a bank with access to the Federal Reserve window during the September 2008 panic. As can be seen, even JP Morgan Chase which is widely seen as relatively prudent under the helm of its CEO Jamie Dimon, was periodically according an elevated probability of default, especially during the financial crisis. Goldman Sachs, whose CEO frequently denied being in need of the Reserve Bank's window, also experienced a sharp increase in implied default by the market (in the right column, we show also the implied equity/debt ratio).

Did the market assign a higher default risk to banks that paid their executives more by high incentive pay? Whereas Fahlenbrach and Stulz (2011) found a neutral to slightly positive effect of incentive pay on default risk for banks, Balachandran *et al.* (2010) used a dynamic panel model and found uniformly that equity-based pay (i.e. restricted stock and options) increases the probability of default, while non-equity pay (i.e. cash bonuses) decreases it.

Of course, banks and financial institutions are different from most other firms for two reasons. Firstly, the failure of financial institutions is often marked by systemic risk which leads to failure of other institutions. Governments have invariably labeled such risk as "too big to fail." This is unfortunate, since studies in the vein of complexity have shown that small banks, even tightly coupled, can also induce systemic risk (Leitner, 2005). The second reason is that because of the threat of systemic risk, the financial services industry, and banking in particular, are heavily regulated. In return, the government provides explicit guarantees that insure

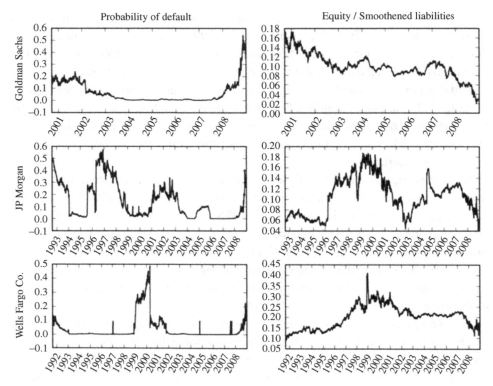

Figure 7.1. Comparison of default probabilities and leverage (market value of equity/smoothed book value of debt). From Balachandran *et al.* (2010).

depositors and implicit guarantees of bailouts for systemically important banks. As a result, both shareholders and managers are incentivized to take on risk, since the government insures the downside.

Returning to the earlier discussion on the Gaussian shape around the regression of log pay on log market value, we can see that the relationship of compensation and market value holds for financial firms also. Why is there more to this relationship than meets the eye, as we claimed earlier? There is good reason to think that this relationship may not represent the assortative matching posited by Gabaix and Landier, but rather, the reward of high market valuation for risky strategies (along with risky compensation policies) when the government insures the downside. It is easy to infer by high market values and high salaries that finance, being complex, attracts "talent," but it is not a priori obvious why finance is more complex than many other industries. This claim is largely unproven. There is a tautological trap in this belief that is often espoused by educated academics, who usually work in finance departments, that goes from high salaries in finance, complex industry, high talent ("smartest guy at the table"), high salaries, ad infinitum. But a simpler hypothesis is that there is an unobservable factor of the "bailout" which inflates both market values and encourages high risk strategies that flow into executive compensation, for a finite period of time.[4] This may be talent of a particular risk inclination and of an irresponsible sort, but not a talent that deserves high pay.

## 7.2  How has the structure of pay differed by industry?

The conditions of government guarantees do not apply to most other industries, except for select cases, such as autos, because of their effects on employment and defense and aircraft production due largely to national security. Consequently, there is considerable heterogeneity across industries resulting from regulatory factors, in addition to technological and competitive reasons. This heterogeneity is revealed in the scale and shape of the size/pay relationships across industries.

In this section and the section to follow, we will provide data in response to questions 2 to 5 listed above. We will not be able to answer why there is such heterogeneity in pay; in our view, this remains a puzzle. However, we will point out several false avenues of investigation and also a few stylized facts that will be useful, we are convinced, for future efforts.

This section establishes that there is considerable variation in pay, which is poorly explained at the industry level by the market value, or size, of the firm. We begin with Box–Whisker plots for the industries in 1992 and 2008 (see Figure 7.2a and 7.2b); they give the mean value within a box whose vertical edges represent the 25th and 75th percentile in pay; the longer lines are bounded by the 5th and 95th percentiles; for three industries in 2008, we have truncated the upper lines for reasons of presentation (all of the values are in thousands and deflated, using the year 2000 price index).

A comparison of the charts for the two years (which are the start and end years of our data taken from Execucomp, part of the Compustat database) shows considerable changes in the leading industries by pay. Firstly, although the data are truncated in 2008, we can see that the largest pay package grew by over 2.5 times in real terms. The high pay industry in 1992 was the food industry – reflecting the primacy of marketing and consumer purchases. In 2008, the top industries were high technology (e.g. software, electronics, chemical), financial services (e.g. insurance and finance companies), oil, and broadcasting. Banking may seem to be a surprising exception; however, the USA has a much larger number of regional banks than most developed countries and these tend to be smaller and, more importantly, 2008 was

---

[4] See Balachandran *et al.* (2010) for a discussion of the evidence that the likelihood of a "bailout" inflates market values.

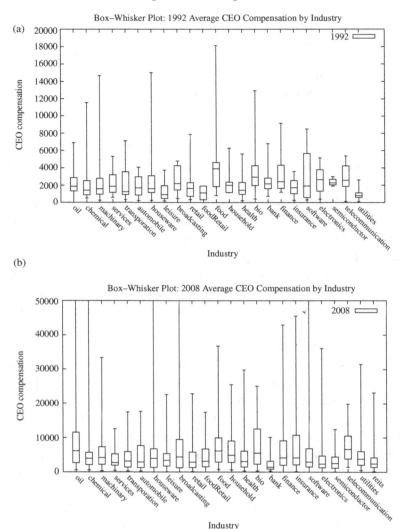

Figure 7.2. Box–Whisker plots of CEO compensation by industry.

in the midst of the financial crisis. Overall, the comparison of these two years shows that pay is much higher on average and the skew of pay appears more prominently.

Another way to look at pay is to calculate the non-parametric Spearman rank correlation using market value and CEO pay; we estimated this correlation for the years 1993 to 2008 with industries. The automobile industry has the highest correlation, followed closely by finance and transportation. At the very end are service firms. Returning to our discussion in Section 7.1, we might suspect that these variations are driven by the *structure* of pay, but in fact, the correlation between these rank correlations (which indicate the strength of the market value size and pay relationship) and pay by equity (stock and options) is only 0.03.

We can be more precise in this discussion of skewness by estimating the relationship of (log) market value and pay through quantile regressions. Quantile regression shows different types of plots by industry,

$$\min_{\beta \in R} \sum_i \rho_\tau(y_i - x_i\beta), \tag{7.1}$$

where, $\tau$ is a quantile, $y_i$ is log *compensation*, $x_i$ is log *market value*, $\rho_\tau(z) = z(\tau - I(z < 0))$, where $I(\cdot)$ denotes the indicator function, and $i$ denotes the $i$th firm. We find the $\beta$ to minimize $\sum_i \rho_\tau(y_i - x_i\beta)$ for given quantile $\tau$. Consequently, $\beta$ is a function of $\tau$, and the following plots are of $\beta$ as a function of $\tau$. We utilize 10 deciles in our estimations.

The surprising results of the quantile regressions shown in Figure 7.3 is the vast sectoral heterogeneity, contrary to the overall pattern in the aggregate data, which shows a reasonably strong relationship between log market value and log pay. For example, a comparison of the first two panels shows that there is a downward slope for automobiles – indicating a falling relationship between the two scales, whereas banks show a moderately rising slope. We can think of reasons why this might be the case, such as the big auto companies dominate the auto industry and its executives capture relatively more pay, whereas banks have many successful smaller players.

However, leaving such speculations aside, the primary conclusion is that there is a lot of heterogeneity in this relationship. In an effort to detect some overall pattern, we divided the industries into two categories by the sign of the regression slopes, giving a value of 1 to industries that show a positive skew, and 0 otherwise. No meaningful patterns were rendered visible by this first-pass effort.

The puzzle here is that the aggregate skewness does not show up in the industry decomposition. The mathematics of power law distributions have the nice property that power law distributed composites (e.g. as firm size by industry) are also power law distributed when added (or even multiplied). It might be expected then that the aggregate skewness of firm size and pay would also hold for a large number of industries. This is not what we find, suggesting that industries are deficient categories by which to understand the micro-dynamics that would drive the overall pattern. This is rather surprising, since industries group close competitors who are utilizing similar technologies and addressing similar markets.

### 7.3 Topology of networks, social comparisons, and pay propagation

An avenue of attack into this puzzle is to consider other influences on pay, besides conventional economic factors, that presumably drive "talent" (i.e. the better CEOs)

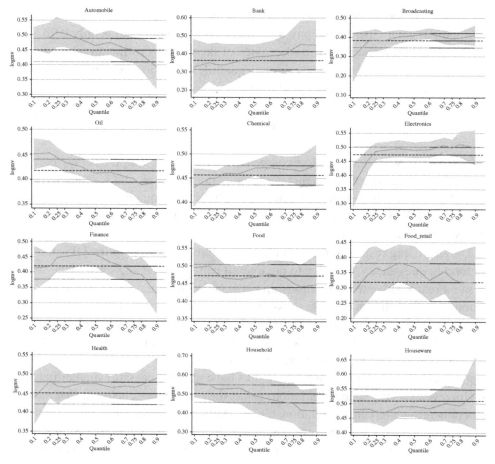

Figure 7.3. Quantile regression of log market value to log compensation by industry. The dashed lines are from ordinary least squares.

to be hired by bigger firms who then pay a premium for their services. One possibility is that pay is influenced by *contagion* caused through a process of social comparisons among CEOs who are connected through social networks.

Figure 7.4 gives the board projection from the board–director bipartite graph for 2008, consisting of 1,851 boards (nodes) and 2,845 board ties (links). The nodes that ring the connected component are the isolates, which make up 32.0% of all firms. There is a giant component in the middle, having an average degree per board of 4.68. While this network is not as connected as one might think, given the many studies on diffusion through board interlocks, there is still a large and well connected component. Only 12.4% of the links are within the same industry.

Since the earlier analysis showed considerable heterogeneity across industries, our initial thought was to redefine the partitions among CEOs/boards by identifying

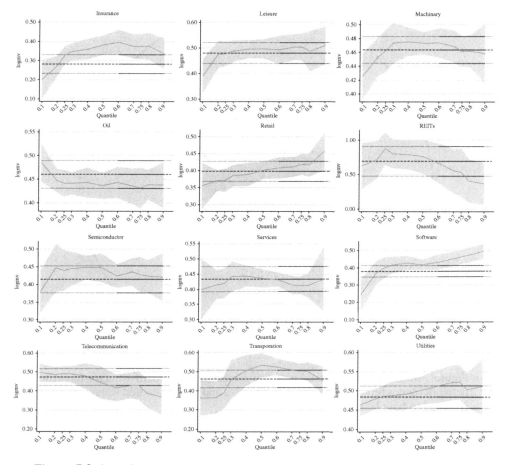

Figure 7.3 *(cont.)*

*modules* that are relatively more connected with each other than with CEOs/boards outside. Modularity is a benefit function to measure how the network is well divided into modules. When modularity has high values, intra-connections between the nodes within the modules are dense while inter-connections between the nodes from different modules are relatively sparse. Modularity ($Q$) is defined as

$$Q = \frac{1}{2L} \sum_{h=1}^{q} \sum_{i,j \in C_h} \left[ a_{ij} - \frac{k_i k_j}{2L} \right], \tag{7.2}$$

where $a_{ij}$ is an element of the adjacency matrix of the network, $k_i$ is degree of a vertex $i$, $L$ is the number of edges in the graph, $i$ and $j$ are $i$th and $j$th nodes from the $h$th group, $q$ is the number of modules (Newman and Girvan, 2004; Newman, 2006).

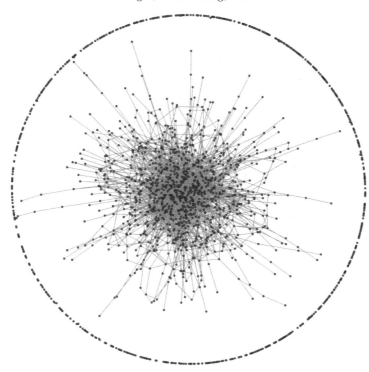

Figure 7.4. Board of directors' network projected from board–director bipartite network (nodes representing board of directors and boards are tied when a director is seated in both boards).

We have used "community detection software" (http://deim.urv.cat/~aarenas/data/welcome.htm) to divide the director networks into modules which make the modularity maximized. After construction of modules, the top 10 and 20 largest modules are used to verify that the CEO compensation levels are different by module. An Anova test indicated that the partitions differ by pay. Even if compensation levels are significantly different by partition, an Anova test for industry shows an even better fit.[5] Thus, modularity analysis did not bring sharper precision into the hidden structure of the director-board topology.

An alternative approach is to treat contagion as between dyadic pairs of CEOs rather than through industry or module partitions. Kim *et al.* (2011) took this approach by testing three kinds of networks: directors, peers, and educational. Two CEOs were matched if they shared a common director, if the CEOs were "peers," or if they were fellow graduates from the same college or graduate school (many

---

[5] F-value = 2.535 for the 10 largest modules, F-value = 1.857 for the 20 largest modules, and F-value = 4.734 for industries.

firms identify "peer" CEOs by which to compare pay and since 2006, the SEC requires all such peers to be identified in filings; we utilized the coefficients from an estimated propensity score model to predict peers for our CEOs). These three types of networks constitute potential effects on the pay of CEOs joined through these dyadic matches. These "network effects" on pay represent a diffusion process by which CEOs, and boards, look at other CEOs and boards to set pay. Previous studies had found that such networks do matter with regard to the level of pay and might explain the heterogeneity we observed earlier. In the Kim *et al.* (2011), we estimated these effects using a logit general estimating equation approach and we also found positive and significant network effects in all three cases.

There is though good reason to fear that network effects are confounded by simple homophily and assortative matching. Good boards will hire good CEOs and pay them well, and directors to good boards will probably be good directors and sit on other good boards that also pay their CEOs well. In this case, the effect is spurious, since the network effect reflects simply that "like" boards share directors, but the determination of pay is not an influence between boards. When we corrected for this type of endogeneity, only the peer effect remained as being clearly significant, with some support for education. Board network influences did not matter.

## 7.4 The effects of mobility on pay

DiPrete *et al.* (2010) call peer benchmarking a "cognitive network," for it is not based on physical affiliation ties, e.g. boards. One can argue whether peer benchmarking is a competitive market, or a sociological effect. It is best in our view to consider benchmarking as both. It offers a possible explanation for why the simple relationship between market value and pay differs among firms: they have different benchmarks.

Peer benchmarking makes sense also because a consequence of social comparison can be that CEOs are hired away or are fired. Our statistical work indicates that hiring an executive from the "outside" (that is, someone who is not already employed at that firm) has a positive effect on pay. At the same time, the longer a CEO is at a firm, the more this tenure decreases the level of pay. If we are to believe that high pay in some industries influences the *general* equilibrium of pay – as proposed in such models as that of Gabaix and Landier – then we are essentially positing a market for CEOs that cuts across industries. How great then is this mobility?

Figure 7.5 presents a weighted graph of mobility for the years 1993 to 2009. The data are for all top executives. The nodes are industries; the size of the node corresponds to the number within intra-industry executive moves; the width of the links between the industries corresponds to the number of executives who moved

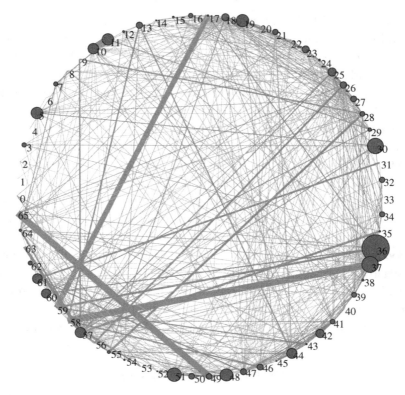

Figure 7.5. Mobility network of executives (nodes are industries and node size is proportional to the frequency of mobility of executives within the industry; width of ties is proportional to the frequency of mobility between industries).

between those two industries; both node size and width are normalized by the number of executives in the originating industry. The largest node is communication; the largest weight on the inter-industry links is between tobacco products and food products. Overall, the percentage of executives changing jobs is 41.3% (1,269) within an industry, and 58.7% (1,805) between industries.

The mobility dynamics in reference to pay are explored in Tables 7.1a, b, and c. Table 7.1a shows very clearly that an important driver of higher pay is the decision to recruit a CEO external to the firm. During the time period 1993 to 2009, about 10.2% of new CEOs were externally recruited and on average they received twice the compensation. We would like to know, of course, if this higher pay is due to CEOs seeking jobs in higher paying industries.

For this, we expand our analysis to the top five executives in every firm, since there would be too few CEOs in each cell to permit analysis. For this reason, the sample sizes are not the same for Table 7.1a compared with Tables 7.1b and 7.1c.

Table 7.1. *Mobility and executive pay.*

(a) Average compensation of new CEO

|          | # sample | Total  | Salary | Bonus | Stock+Option |
|----------|----------|--------|--------|-------|--------------|
| Insider  | 4486     | 3438.0 | 533.7  | 484.5 | 1650.0       |
| Outsider | 508      | 8176.5 | 631.4  | 749.2 | 5114.7       |

(b) Average differences of executives' compensation when they move to another industry

|       | # sample | Total  | Salary | Bonus | Stock+Option |
|-------|----------|--------|--------|-------|--------------|
| 1990s | 672      | 3121.6 | −43.8  | 161.7 | 3013.2       |
| 2000s | 1133     | 226.0  | −59.4  | −60.7 | −192.8       |

(c) Difference of the average executive compensation between the current and the previous industries when executives move to another industry

|       | # sample | Total  | Salary | Bonus | Stock+Option |
|-------|----------|--------|--------|-------|--------------|
| 1990s | 672      | −100.7 | −7.9   | −32.2 | −49.3        |
| 2000s | 1133     | 24.6   | −4.6   | 18.8  | −0.2         |

Table 7.1b indicates that for the individual executive, the effect of accepting a job in a new industry in the 1990s was very beneficial, largely due to the very high compensation in stock and options. The increase in stock and option compensation reflects both the increased value of the stocks during the Internet boom, as well as presumably the enticements to attract executives during this time. Not surprisingly then, the bust of the Internet boom led to a fall in compensation levels overall (as we know from the general time series of pay). Thus, executive mobility resulted in a much lower increase in total compensation (it should be noted that total compensation also includes pension benefits, which are not shown). There was then a correction for the boom salaries of the Internet gold rush.

An interesting comparison is provided by Table 7.1c which looks at whether executives were changing jobs for higher paying industries; we can identify high and low pay industries by calculating the average compensation in an industry. The statistics show that executives were not moving to the higher paid industries in the 1990s, but by inference from Table 7.1b, they were choosing the high paying jobs in lower paid industries. This is a very clean insight into the diffusion dynamics: executives were migrating from high paying industries to high paying jobs in lower paying industries. By the 2000s, executive were willing to move without larger increases but into higher paying industries.

In all, there is a pattern of the equilibration of pay across industries. These movements show a global market for executives who are seeking better paying jobs, while the precise industry choices are very much affected by the general stock market. Nevertheless, it is important to remember that almost 90% of new CEOs are internal to the firm. Thus, the role of the external market is to determine the marginal pay offer, which through a process of peer comparison can then percolate to other firms and industries.

## 7.5 Local networks and gender

The puzzle we are addressing is why there is so much heterogeneity in the pay for CEOs and executives. We have studied above the possibilities of diffusion through director networks and have summarized the work on contagion by other networks, especially through peer benchmarking. The last section found that mobility during the Internet boom was an important source for rising CEO pay, which was partly corrected during the subsequent decade, starting in 2001.

Another potential source of heterogeneity is "negative homophily," whereby there is a prejudice against particular types of executives and CEOs. Given the large amount of evidence on gender effects on pay, an obvious candidate is the potential inequality among male and female executives regarding pay. A major obstacle in studying gender prejudice is that between 2 to 3% of CEOs are women, though the percentage of female directors is larger. For female CEOs, boards have about 25% female directors, against 10% when the CEO is a male. There is clearly assortative matching between the gender of the CEO and the percentage of women on a board.

Nevertheless, a simple test shows that female and male CEOs are paid approximately the same for controlling industry. Suggestive evidence for gender bias is found, though, when we examine the changes in pay when a new CEO is hired. Table 7.2 reports the statistics for increases in pay when the CEO changes by gender. Keeping in mind the small sample size, the striking comparison is between rows two and three. When a woman replaces a man as CEO, the mean compensation increases by 0.45. However, in the case of a male replacing a female CEO, the mean pay increase is by a factor of 3.38 (note with caution the large standard error). The suggestion is that gender matters to pay dynamics when there is a change in CEO.

One reason why there is no clear pay difference is because more women are found in some industries than in others. Particular industries have more female directors than others, such as household products, food and retail, and utilities. Since we know from the above discussion that larger female representation on a board leads to a greater probability of hiring a female CEO, it is not surprising that there is also an industry pattern to female CEO employment. Women CEOs and women directors tend to be found in the lower paying industries.

Table 7.2. *Growth rate of CEO compensation when CEO changes by gender of the previous and current CEOs.*

|                  | Sample size | Mean | Std err |
|------------------|-------------|------|---------|
| Male to Male     | 1,403       | 1.17 | 0.139   |
| Male to Female   | 22          | 0.45 | 0.293   |
| Female to Male   | 7           | 3.38 | 2.218   |
| Female to Female | 2           | 0.90 | 0.571   |

## 7.6 Conclusions

There have been many studies on executive pay but there remains a remarkable lack of knowledge concerning the determinants of the level of pay: what is too little or too much and why is there such heterogeneity? We have noted some of the regulatory issues, such as the restriction on board interlocks for banks, which can affect the pattern of contagion across industries. There are also the effects of social networks due to shared educational ties or assortative matching due to gender homophily.

The broad sweep of our analysis points to the importance of mobility for two reasons. The first is that mobility increases during periods of boom and bust. During booms, CEOs move to higher paying jobs, even if located in lower paying industries. This mobility is motivated by the very large heterogeneity in pay. A Gini coefficient is a measure of inequality bounded by [0,1]. This coefficient moved from 0.25 in the mid 1990s to 0.4 by 2000 and 2001, falling back to 0.25 by 2006, and then even further to 0.2 in 2008, when market values, and the value of stocks and stock options, collapsed. These two observations of high mobility to higher payer jobs and high variance in pay are clear signatures of diffusion. Secondly, during busts, CEOs are fired, leading to higher vacancy rates and mobility, although pay does not increase. There is therefore also a dampening effect to the positive feedback caused by mobility through the general depression of market valuations and therefore on executive pay.

Following the results of Kim *et al.* (2011), we doubt very much that contagion is channeled very strongly by board–director ties. Instead, the evidence points to peer benchmarking, which has been shown to be partly determined by observing past movements of CEOs between firms (Faulkender and Yang, 2010). Nevertheless, the cumulative effect of network ties may be larger than suggested by dyadic comparisons, for two reasons. Firstly, CEOs are connected by many kinds of networks, such as educational, peer, social, and board interlocks; in their joint effects, these influences may contribute substantially to diffusing pay among firms. Secondly, network influence is often based on average peer effects or dyadic models. Whereas these effects may be small, they can be enlarged when multiplied by the CEO network degree. Thus, for reasons of multiplexed networks and the average degree of a CEO, the cumulative and joint effect of social influences on executive pay is potentially large, even if elusive to measure. Therefore, to answer more fully why we observe particular levels of pay will require dynamic modeling of multiplex networks and CEO mobility. Our analysis points in this direction.

# 8

# Networks of institutional capture: a case of business in the State apparatus

EMMANUEL LAZEGA AND LISE MOUNIER

Businesses of all kinds usually try to participate in the regulation of their own activities as much as they can. One way of participating in this regulatory activity is to exercise control on State institutions which solve conflicts among businesses and discipline entrepreneurs. This participation can lead to institutional capture. In this chapter we describe and show the contribution of network analysis in measuring the level of institutional capture in a specific case. We analyze a network of business people acting as lay judges in a judicial institution, the main first-level commercial court in France. This court handles commercial litigation as well as bankruptcies. Courts are not static institutions making atemporal and purely rational decisions. They are contested terrain, the object of broader conflicts that occur outside courthouses. We use a longitudinal study of advice networks among these 240 lay judges at the Commercial Court of Paris (CCP) – judges who are elected by their business community at the local chamber of commerce – to examine a few characteristics of such an institutional capture, in particular an invisible mechanism through which the banking industry manages to dominate this court. Results illustrate the value of network studies for a renewed attention to the inner workings of institutions and for the protection of public interest in the regulation of capitalist economies, in which boundaries between private and public sectors are blurred.

## 8.1 Joint governance and institutional capture

Business usually tries as much as it can to participate in the governance of its markets. In this chapter, we look at an extreme case of how business gets organized, collectively, to do so by capturing a judicial institution. The case is that of French commercial courts, a four and a half century-old institution. In France, the State

Networks in Social Policy Problems, eds. Balázs Vedres and Marco Scotti. Published by Cambridge University Press © Cambridge University Press 2012.

has long been sharing its own judiciary power with the local business community. Business succeeded in 1563 in negotiating what could be called a "joint governance" agreement with public authorities: this agreement created a special jurisdiction for commerce in which judges are lay, voluntary judges, i.e. elected business people who are not remunerated for the job. French commercial courts are truly judicial, first-level courts. They solve conflicts between businesses or between businesses and consumers (commercial litigation). They also exercise a form of discipline on market exit by handling bankruptcies. Their capture is the result of a complex historical and institutional process. Here we focus on the dimension of this process that is brought to light by social network analysis.

Institutional capture can be defined as "the efforts of firms to shape the laws, policies, and regulation of the State to their own advantage by providing illicit private gains to public officials" (Hellman and Kaufmann, 2001; Kaufmann and Kraay, 2007). We suggest that this definition is too focused on individuals. We think that the definition of the process of institutional capture should be broadened to involve corporatist efforts to design or redesign institutions, to influence decision making in rule enforcement, and to obtain collective gains for interest groups in these institutions. These elements add to the capacity of collective actors to gather invisible advantages. A court can thus be captured, inasmuch as interest groups are successful in using their influence to benefit systematically from its decisions.

This institution represents a form of joint governance or a combined regime of endogenous and exogenous conflict resolution in markets. We use the label "joint" because we argue that governance is often a combination of self-regulation and exogenous regulation, a combination in which costs of control are shared. We can define the joint element in "joint governance" as the coexistence of several sources of constraint, both external and internal, that weigh on the actors in charge of solving the conflicts and enforcing the rules. Thinking about joint governance in these terms follows both an organizational and a broadly conceived structural approach to economic institutions (Lazega and Mounier, 2002; Lazega, 2009). Courts are indeed a locus of joint governance. They are not static institutions making atemporal and purely rational decisions (Heydebrand and Seron, 1990; Wheeler *et al.*, 1988). They are contested terrain, the prizes or objects of broader economic competition and conflicts that occur outside courthouses (Flemming, 1998). This is especially the case in courts in which judges are themselves business people elected by their local business community. Attempts at influencing what goes on in court from outside the court come through various angles. Flemming (1998) lists five such angles: external stakeholders can try to influence jurisdiction (the range of disputes over which the court has authority), positions (actors formally authorized to participate in the disposition of cases), resources (the capacity to influence the decisions of other actors), discretion (the range of choices available to actors), and procedures (rules governing courtroom processes). Parties involved in that contest may not be directly concerned by all the conflicts that are dealt with by the court, but they may have indirect concerns, material or symbolic, in the decisions of the court, and thus attempt to influence what goes on.

Flemming's categorizations help focus on specific processes of influence. We limit ourselves to the study of two such processes: position and resources (in Flemming's vocabulary) and the relationship between the forms of influence they represent. By this we mean that we look at who is allowed to become a judge and what kind of resources are made available to them when sitting in judgment and participating in governance while solving conflicts between businesses.

Indeed, collective actors involved in conflicts on markets, such as companies, whole industries (in class actions, for example), and even State administrations, may have strong incentives to influence who becomes a judge and what resources are available to these judges. For example, the more litigious these sectors are, the stronger their incentives to share the costs of conflict resolution. These collective actors are usually conceived of as external actors. Their incentives to influence the jurisdiction of the court are to protect their own interests in the long run. They may do this by helping some of their own selected members in becoming a judge. The stronger their incentives, the more they can be expected to take care of their representation among the judges. Such judges, once in a position to solve conflicts between parties, may have combined incentives to influence the jurisdiction of the court: on the one hand, they represent the law and are meant to proceed in all independence; on the other hand, they may represent, and therefore protect the interests of, the organizations that supported their becoming a judge in the first place.

Influencing who becomes a judge and what resources are available to these judges can be a very strong, although indirect, way of influencing the outcome of judicial decision making. In effect, collegial deliberation among judges is built into legal institutions. This deliberation – whether formal or informal – relies heavily on knowledge management by the court. This brings us to the second process (in Flemming's list above) by which influence characterizing joint governance is exercised in such organizations. One way to influence the behavior of judges is to try to set the premises of their decisions by attempting to control the information available to them while sitting in judgment. Judges coming from one sector of the economy may thus, as "judicial entrepreneurs" (McIntosh and Cates, 1997), attempt to keep particular definitions of problems alive, or promote the ideas, customs, rules, and interests that are commonplace in their sector, although not in others. Control over the premises of decisions can be assumed to influence the probability of who wins, even if this "framing control" by players is almost invisible to outside observers.

The law and the courts are aware of the fact that different types of actors in the environment of the court are involved in such influence attempts. Anticipating that the court will be the target of such external attempts at influence, the legal system provides rules concerning conflicts of interests for judges: when they are too close to one of the parties, for example, when they sit in judgment of a competitor, sometimes even of a business in the same industry as their industry of origin, they must self-disqualify or, if this is discovered, be removed from the case by their hierarchy.

However, a structural approach to joint governance raises the issue of the success of such procedural attempts in neutralizing external influences, especially when the judges are elected. Indeed, and to summarize, given the incentives identified above, we can expect influence on judges to take a form identified in the following hypotheses. Firstly, the stronger the external stakeholders' interests and expectations, the more likely they are to invest in shaping the courthouse, especially in selecting the judges. Secondly, the closer the judges (as levers of influence) are to other judges (as targets of influence) in the same court, the more influence they have on these other judges (i.e. on their targets). In other words, being active in the social life of the courthouse multiplies opportunities to meet, discuss decisions, and therefore of being sought out for advice, thus increasing one's influence on other judges as targets.

In this chapter we test these hypotheses using a network study of the Commercial Court of Paris (CCP). Network data are used to show the existence and the determinants of this influence structure.

## 8.2 Consular commercial courts as joint governance institutions

Consular courts are operated by consular judges. Each consular judge acts both as an individual judge, and as a representative (presumably without any specific mandate) of the business community. Consular judges are voluntary, unpaid, and elected for two or four years (for a maximum total of fourteen years) at the Chamber of Commerce of their local jurisdiction. Judges are elected after a complex procedure (Falconi *et al.*, 2005) that begins with individually obtaining the sponsorship and approval of a professional association (for example the French Banking Association, the French Hotel Industry Association, etc.). The electoral body is composed of currently sitting judges at the commercial court and of representatives of the employers' associations (some of whom sponsored the candidates in the first place). A small administrative unit of the Chamber of Commerce searches for new candidates, interviews and selects them, composes a unique list of candidates that is then submitted to the vote of the electoral body for formal rubber stamping. Although no part of a democratic government is usually allowed to self-perpetuate in that way, consular judges have thus co-opted each other for centuries.

In this institutional solution, the State, the industries or companies, and the individual judges share the costs of exercising social control of business. These judges sit in judgment one day a week to enforce law and customs among their peers. Decisions made by the court can be challenged, as in any court, and be brought to the Court of Appeal. At the Court of Appeal, judges are no longer business people, but highly selected professional judges. In such a system, there are at least two categories of such unpaid, voluntary, and consular judges: firstly, retired business people looking for social status, an interesting activity and social integration; and secondly, younger professionals, whether bankers, lawyers, or consultants, who look for experience, status, and social contacts, sometimes on behalf of their employer (who keeps paying their salary while they are practising as a judge at the Court). If the individual judge is young enough, this can help to build a relational capital (as explicitly stated in the flyers trying to attract new judges to the job) and to open doors for future positions in the economic institutions such as the *Chamber of Commerce* itself, sometimes the *Conseil Economique et Social* (an advisory board to the Prime Minister), and other honors dispensed by the State apparatus. Indeed, for younger professionals, sitting as a judge at the CCP has traditionally been considered a "chore" that would be rewarded later on with seats on prestigious committees in the economic institutions of the country. Various types of lucrative contracts and missions may also be awarded, on a discretionary basis, by the acting President of the Commercial Court to former judges to advise companies on a "prevention" basis. Consular judges can also become arbitrators in arbitration courts, once they are done with their 14 years at the public courthouse.

There are several justifications for this joint governance regime. Firstly, it is a cheaper and faster form of justice than a system with career judges. Business bears more of the costs of its own regulation, backlogs are much smaller, and waiting time is shorter than in traditional High Courts. For example, there is no first-level jurisprudence. Secondly, career judges – who are civil servants – have often been considered to be inexperienced in business and unable to understand the problems of private companies, or to monitor satisfactorily the behavior of company directors, particularly in the insolvency and bankruptcy minefields (Carruthers and Halliday, 1998). Thirdly, business law often ignores idiosyncratic norms and customs (called *usages* in French commercial courts) based on industry traditional subcultures characterizing whole sectors (Macaulay, 1963; Swedberg, 1993). The argument is that efficient conflict resolution cannot ignore these bodies of rules and conventions, that organize business practice differently in each traditional sector. Judges at the *Tribunal de Commerce*, since they are supposed to be experienced business people, are thus said to be specialized in their professional field, and in a better position than tenured civil servants to know about these customs; to adjust them more quickly to unstable or changing business environments; and to be in a better position to foster regulatory innovations, either as experts in their field consulted by Parliament, or as members of think tanks.

In this consular system, the difficulty with representing general and particularistic interests has always been the predicament of these judges. For example, when they are elected from among business people, they can be thought of as representing the State as well as their business community. They may claim that they are not "representatives" with a mandate from the industry that helped them become a judge. But they are still expected to speak on behalf of this industry and its customs by the members of this industry, and sometimes by fellow judges. The public has always suspected that patronage appointments lead to politicized elections of judges, who then fail to distance themselves from their virtual "constituency," i.e. the industry that endorsed their candidacy for the job. Especially in small town commercial courts, litigants' confidence in the impartiality of the tribunal's decision is often impaired. They fear that judicial control can be exercised by competitors. The institution, however, assumes that zealous judges will be entirely virtuous, in spite of the short distance between regulator and regulatee.

### 8.3 Networks at the Commercial Court of Paris

Fieldwork was conducted at the CCP. This court is one of the four large commercial courts in the Paris region. It includes twenty-one generalist and specialized chambers (such as bankruptcy, unfair competition, company law, European community law, international law, multimedia, and new technologies, etc.) which handle around 12% of all the commercial litigation in France, including large and complex cases (when they do not go to arbitration courts). There are around 150 judges each year. Socio-demographic characteristics of the judges show that their average age is 59; 87% of the judges are men; 38% are retired. Positions occupied by judges in their industry and occupation of origin (or former occupation) include CEOs (25%), vice-presidents, and top executives of all kinds. Amongst the youngest judges, there are more professionals such as company counsels, accountants, and consultants. Most of the time, they work for large business groups or medium-sized companies (that the judges do not like to name); they prefer to be discreet about their professional affiliation. Most judges are highly educated with diplomas from top French business schools, law schools, elite engineering schools (called *'noblesse d'Etat'*).[1]

Fieldwork was carried out over five years. Response rate reached 90% on average at each phase of the study. A first phase took place in 1999, during which we collected socio-demographic information about the judges, ethnographic information, and

---

[1] More details about this institution are provided in Lazega and Mounier (2003).

observations on the operations of the court. A second phase in 2000 allowed us to interview face to face all the judges, about various issues of interest to the presidency of the court, and to include a name generator about advice seeking in the questionnaire. During the third phase, in 2002, we interviewed all the judges about their motivations, careers, and values, plus a second measurement of the advice network among the judges. Finally, a fourth phase in 2005 allowed us to collect systematic materials on the judges' jurisprudential reasoning (using vignettes and real-life court cases), as well as a third measurement of their advice network. The organizational functioning of the CCP is complex. It is not our purpose to describe it here in more detail. As mentioned above, several kinds of professionals operate these courts together: consular judges, clerks, business lawyers, representatives of the attorney general, bailiffs, experts of all kind, professional liquidators and/or "administrators" (for companies that are on the brink of bankruptcy but could perhaps be brought back to life). Judges are allocated across the large number of generalist and specialized chambers. The minimal distinction in terms of specialties is between bankruptcy and litigation. Different procedural rules are legally attached to each. But the litigation bench is then subdivided in many types of subspecialties mentioned above. There is a president of each chamber who reports to the president of the court. In each chamber, three – sometimes five – judges sit together in judgment, processing cases collegially while listening to the conflicting parties, as in any other judicial court.

### 8.3.1 Networks in the recruitment of consular judges

Recruitment of judges is always a difficult task. As seen above, recruitment of consular judges adds a level of complexity; it is severely controlled by the business community and its institutions. Candidates for the position are often initially contacted by sitting judges, colleagues, professional associations, or even by family members who are themselves part of this relatively closed consular milieu. Here we were able to elicit from the judges information that shows that social networks matter in the recruitment of future candidates. We asked the judges if they had had informal discussions about the idea of applying for the position before they became a judge themselves, and if so with whom. Eighty percent declared that they had had such informal discussions. In 57.4% of the cases they mentioned discussions with judges already sitting in the Court. Consular judges are thus the main recruiters of new consular judges. They play the main role in attracting new candidates using their informal relationships. Their relational work is crucial in the generally difficult task of recruiting new magistrates upstream of elections and formal co-optation.

### 8.3.2 *Over-representation of the financial industry among lay judges*

According to the justification of this system of joint governance, the selection of judges should produce a representation including as many sectors of the local economy as possible, especially in large commercial courts such as that of Paris. At the time of the study, economic sectors represented by the judges (i.e. in which they work or in which they used to work) were indeed very diverse. Thus, in complex cases, intelligence about a sector could be made available to the court from the judges coming from that sector. However, some industries or companies invest more than others in judicial entrepreneurship, and incur a larger share of the costs of control because they have an interest in doing so. Although in theory all employers' associations can present candidates to the elections of consular judges on an annual basis, to fill in the vacancies created by a 10% turnover at the court, in reality all do not and some do so much more systematically than others. In 2000, 29% of the judges came from the financial industry. Forty-four consular judges, the majority of them with a legal education, were currently employed by the financial industry or had been employed by it in the past. This industry promotes several candidates for election to the job each year.[2] The financial industry is clearly over-represented, in absolute and relative terms, at the Commercial Court of Paris. In effect, the sector of financial activities represents around 5% of the active population in Paris where the services industries are over-represented compared to the rest of France. In terms of value added to the economy per branch, the share of the financial services industry in the total value added to the French economy was on average 5.3% at the time.[3]

This industry is traditionally a very litigious sector (Cheit and Gersen, 2000). The business docket in France, probably as in most countries, is dominated by contract disputes and debt collection issues. A sizable portion of this docket involves the financial industry, for obvious reasons, which provides a strong incentive to invest in judicial entrepreneurship – for example to ensure damage control in cases involving high levels of credit. Banks and financial institutions are often creditors themselves and they stand to lose enormous amounts. Therefore they invest in penetrating the courts and maintain a high number of consular judges coming from their ranks. The financial industry, with its high amounts of resources at stake in commercial litigation and bankruptcy, is willing to play for the rules. It has an interest in trying to shape the court and impose its industry norms and practices over those of other industries. The priorities of the financial sector (such as preserving the high value of

---

[2] For example, 21 were elected as candidates of the *Association française de banque* and five as candidates of the *Association française de sociétés financières*. Among the financial companies that were the employers of sitting judges (at the CCP alone), BNP-Paribas had sent seven judges, Suez had sent four, Société Générale four, Crédit Lyonnais four, and Crédit Commercial de France four.

[3] *Source:* Institut national de la statistique et des études économiques, Comptabilité nationale, 2001 (http://www.insee.fr).

assets, high level of claims, and high sensitivity to the impact of corporate failures on the economy) can thus be defended in both the litigation and the bankruptcy bench. One of the likely influence processes that characterize joint governance is thus detected in the selection of judges themselves (i.e. the "positional" effect in Flemming's vocabulary). Small employers' associations do not have the necessary clout to lobby effectively, nor enough resources to share the costs of control. All sectors of the business community cannot participate equally actively in the contests and attempt to shape the courthouse from outside. The potential influence of each in the struggle over this kind of contested terrain varies with the resources available to promote candidates for the jobs of consular judge. Resources available are not the same in the banking industry and in less well organized sectors, such as retail.

One of the combined effects of the fact that consular judges are the main recruiters of consular judges, and of this over-representation of bankers, is that bankers, being more numerous, also attract an increasing number of bankers to the court. These descriptive results also show that judges coming from the financial industry are clearly potential levers of that industry. Influence exercised by these judges over other judges in the court would mean that the financial industry is the biggest threat to this court's independence. In effect this over-representation allows the banking industry to be over-represented in key chambers. For example, in 2000, bankers were over-represented in bankruptcy chambers. This over-representation reaches a peak in the chamber hearing appeals against decisions made by the bankruptcy bench (a sort of internal Court of Appeal for decisions about bankruptcies): in this chamber, five out of seven judges were bankers, raising clear issues of conflicts of interest.

### 8.3.3 Advice networks and bankers' influence

Using this information about "positional" influence, we can now look at whether or not judges coming from this over-represented financial industry are in a position to exercise invisible influence on other judges, by providing them with resources such as information and advice. We measure this influence by looking at centrality in the advice network amongst all the judges of the court. We assume that advice interactions between judges are equivalent to interactions setting the premises of judicial decisions, especially since bankers with a law degree may be sought out for advice, because they have a better legal education than other lay judges. In effect, patterns of advice seeking in the court show who is prepared to listen to whom when framing and defining problems at hand in the judicial decision making process. The advice network among judges can thus be considered to be a bridge between structure and decision making (Lazega, 1992). Therefore, we look at the ways in which the judges transfer and exchange advice, then try to appreciate the ways in which contextual factors can influence this process. We can measure the capacity of an industry to set the premises of such decisions by looking at the centrality of its representatives in the advice network amongst all the judges, and then at the determinants of this centrality.

Data on advice seeking among judges was collected in 2000, 2002, and 2005 using the following name generator: *"Here is the list of all your colleagues at this Tribunal, including the President and Vice-Presidents of the Tribunal, the Presidents of the Chambers, the judges, and 'wisemen'. Using this list, could you check the names of colleagues whom you have asked for advice during the last two years concerning a complex case, or with whom you have had basic discussions, outside formal deliberations, in order to get a different point of view on this case?"* Our very high response rate allowed us to reconstitute the complete advice network (outside formal deliberations) amongst judges at this courthouse, three times as a longitudinal dataset, and thus to measure each judge's centrality in this network over time.

As in advice networks more generally (Krackhardt, 1990), an informal pecking order or status hierarchy emerges amongst judges. In order to look at the relationship between bankers and centrality, we included these attributes along with several other characteristics of the judges in a regression model predicting centrality in the advice network amongst judges. In addition to the main variable representing the judges sector of origin (having worked in the financial industry, combined here with holding a law degree), a series of control variables were added to the model. Seniority, measured by the number of years an individual has served as a judge, can be understood as "experience" and help a judge to wield influence, independent of the sector of origin. In addition, the other judges in the commercial court may not be the single source of advice and influence. Being well connected and open to the business community can attract colleagues who need economic advice. The same is true with being well connected and open to professional judges in other courts (especially the Court of Appeal, where judges are career judges). Such external ties can attract colleagues who need legal advice. The same could happen with being well connected and open to the office of the Attorney General, although surveillance and influence by the Ministry are not always welcome in consular courts. Being in activity (as opposed to retired) may also have an effect on centrality in the advice network, since retired judges may have more time and be more available than professionally active ones to discuss issues at length. Being a member of the State elite (the *noblesse d'Etat*, i.e. coming from the elite French engineering and administration *grandes écoles*, having been trained at elites schools such as *ENA* or *Polytechnique*) means that one has connections in high places, and thus might wield some authority and influence among fellow consular judges. Table 8.1 presents the analysis controlling for these effects.

Table 8.1. *Key (most central) players in the advice networks among lay judges at the Commercial Court of Paris in 2000, 2002, and 2005. Linear regression model measuring the effect of lay judges' characteristics on their centrality in the advice network. The test we use is the t-statistics, defined as estimate (parameters) divided by standard error (S.E.). When this t-test is greater than 2 in absolute value, the parameter is considered significant at the 0.05 significance level.*

| | 2000 | | 2002 | | 2005 | |
|---|---|---|---|---|---|---|
| | Parameters | S.E. | Parameters | S.E. | Parameters | S.E. |
| Intercept | −3.54 | 1.02 | −1.11 | 1.65 | 1.08 | 1.61 |
| Seniority | 0.67 | 0.08 | 0.80 | 0.12 | 0.72 | 0.13 |
| *Noblesse d'Etat* | 1.13 | 0.90 | 3.04 | 1.42 | 1.67 | 1.57 |
| Professionally active (versus retired) | −0.61 | 0.63 | 0.12 | 0.92 | −0.26 | 1.02 |
| Bankers with law degree | 1.33 | 0.71 | 2.93 | 1.09 | 3.14 | 1.32 |
| Participation in social functions | 2.36 | 0.92 | 0.23 | 1.30 | 1.80 | 1.31 |
| Seeks advice: | | | | | | |
| *–from business sector* | 1.61 | 0.62 | 0.05 | 0.92 | −1.43 | 1.14 |
| *–from career judges* (CoA) | 4.49 | 1.42 | 5.09 | 1.93 | 2.56 | 1.85 |
| *–from district attorney* | −1.72 | 0.63 | −1.70 | 1.12 | −0.25 | 1.22 |

Results expose the informal and indirect influence of bankers with a law degree over their fellow consular judges. The sector of origin of a judge, particularly coming from the banking industry and having a law degree, has a significant effect on being central in almost all three models. Bankers are over-represented at this court, and among them, bankers with a law degree exercise strong indirect influence through premise setting in this organization.

Controlling for the other variables, being active in the social life of the court has an instable effect (significant in one model only) on centrality in the advice network, and thus on the capacity to set the premises of other judges' decisions. In effect, in order to exercise such indirect influence, judges should also be strongly integrated in the courthouse, in its social life, have and use ties outside the courthouse, consult with professional judges. Notice that, in addition to being socially active in the court and to being a banker with a law degree, being senior and seeking advice from other sources (in the business community and among professional judges) are also good predictors of potential influence in this court. Notoriety of consular judges can be built inside the small microcosm of the courthouse by investments in ties to other judges, whether within or outside this specific courthouse. Notice that seeking advice from the Attorney General (who represents the State directly inside the Commercial Court of Paris) is significant and negative in 2000 (under a Socialist government): the more contact judges have with the office of the Attorney General and its representatives, the less these judges are sought out for advice by their peers. In effect, the parameter estimate is longer than twice the standard error in 2000 (socialist government), but no longer in 2002 and 2005 (Conservative government). In summary, the more socially active the judges within the court, the more open to discussions with the business community and with the legal environment – but the less open to discussions with representatives of the State – the more influential these judges are at the court.

Finally, it should be stressed that bankers' influence, particularly when they have a law degree, has consequences in terms of decision making. For example, bankers are mostly non-punitive (Lazega *et al.*, *in press*) i.e. less keen on awarding "moral" damages to plaintiffs, in cases of unfair competition.

## 8.4  Conclusions

With increasing porosity of the boundaries between the private sector and public institutions in advanced capitalist societies, institutional capture in joint forms of governance has become a policy issue even in domains that have usually been con- sidered closer to the core functions of the State, such as education, health, family, security, or science. Private economic actors in these market areas spend time and resources trying to organize their environment, to improve their opportunity struc- ture, and manage the governance mechanisms that constrain them. These efforts are often built into the operations of economic institutions, especially institutions representing joint governance. We show here how network analysis is efficient at measuring a level of institutional capture that is usually difficult to observe in complex joint governance by State and private actors (Lazega, 2003). In our case study, State captors can be representatives of the oldest incumbents, not new market entrants as in Stark and Bruszt (1998) or Hellman *et al.* (2000).

Re-examining the institutional frameworks of market governance using network analysis can shed light on the mechanisms that facilitate institutional capture. In our case study, a complex system of cooperation between the State, local Cham- bers of Commerce, and citizens, produces consular commercial courts and specific ways of sharing the costs of social control of business. In particular, we focus on the regulatory influence and special role of the financial sector in this process: when business gets organized collectively to connect to the public sector, the dual character of this financial sector (both economic and political) and its regulatory role and influence can be brought to light. In our case study, the importance of the financial industry is measured not only by the number of consular judges that it places on the bench, but also by the centrality of its representatives in the advice network of the organization. This epistemic influence at the implementation phase of market governance provides a level of control of the institution that is difficult to acknowledge and measure without knowledge of social networks.

As joint governance increases, so does – in our view – the danger of widespread institutional capture. Our study suggests that public policy-makers could benefit more systematically from organizational and structural studies of joint public–private governance arrangements by looking at such institutional arrangements through this lens. We suggest that this approach can be used in the future to rethink the notion of conflict of interests (Lazega, 1994, 2011), a dimension of relational embeddedness of economic action that has been relatively neglected both in the scientific and policy literature. Social and organizational network analysis can be very effective in detecting situations of conflicts of interest and institutional (not necessarily personal) corruption. Combined with organizational analysis, it can be an efficient method of measuring the level of capture or of independence of public office in such complex situations.

# 9

# The social and institutional structure of corruption: some typical network configurations of corruption transactions in Hungary

ZOLTÁN SZÁNTÓ, ISTVÁN JÁNOS TÓTH, AND SZABOLCS VARGA

In the first part of the chapter, four ideal-typical corruption transactions are explicated in terms of the principal–agent–client model: bribery and extortion are described as two different types of agent–client relationship, while embezzlement and fraud, as two different types of principal–agent relationship. The main idea is to describe these elementary corruption transactions as simple directed graphs. The next section of the chapter takes into consideration different kinds of possible motivation (such as the reduction of risk or transaction costs) of the principals, agents, and clients, in order to embed their corruption transactions in various kinds of personal, business, political, and other institutional networks.

In the second part of the chapter some typical and stable network configurations are presented, based on recent empirical corruption research carried out in Hungary. Certain corruption cases (such as party financing or granting of permits) are analyzed in detail, and are described as complex and multiple networks. The chapter concludes by showing some signs of the evolution of corruption networks in Hungary in terms of the number of actors, the complexity of network configurations, the level of personal or institutional embeddedness, and the multiplicity of relationships.

## 9.1 Introduction[1]

The study consists of two main parts. In the first part, the concept and ideal-types of corruption are defined. The various types of elemental corruption transactions

[1] The study was done at the Corruption Research Center of the Institute of Sociology and Social Policy at Corvinus University of Budapest. The research which served as the basis of the study was conducted by: university students Hilda Kinga Balázs, Kinga Bartis, Tünde Cserpes, Márk Tamás Fülöp, Gergely Lukácsházi, Annamária Márkus, Erna Miskolczi, and Orsolya Vajda, as well as doctoral student Szabolcs Varga. The research was led by university professor Zoltán Szántó and senior research fellow István János Tóth. The research was supported by TÁMOP 4.2.1.B-09/1/KMR-2010-0005 project.

Networks in Social Policy Problems, eds. Balázs Vedres and Marco Scotti. Published by Cambridge University Press © Cambridge University Press 2012.

are differentiated in terms of the principal–agent–client model, illustrating them through directed graphs. We distinguish two subtypes of both the agent–client relationship and the principal–agent relationship: bribery and extortion in the former case, embezzlement and fraud in the latter. To conclude the first part, we attempt to delineate the motivational mechanisms that encourage participants in corruption scenarios to embed their transactions in various types of personal, business, political, and other institutional networks. With the help of these networks, those involved are often able to decrease the transaction costs and risks associated with corrupt dealings.

In the second part of this chapter, the social and institutional embeddedness of a few typical Hungarian corruption transactions is discussed and illustrated through multiplayer, complicated, multiplex graphs. Here, we will supplement the ideal-typical models of corruption with the concepts of the hidden principal, the broker, and the hidden role. Through the latter, we wish to demonstrate that those involved in corrupt transaction may, in certain scenarios, behave differently than would be expected based on their roles. Our previous research has shown that, in corruption transactions, the most common corruption risks are associated with permit/license acquisition and inspection, the acquisition and use of EU funds, public procurement, as well as the buying and selling of government and local government-owned real estate (Alexa *et al.*, 2008; Szántó *et al.*, 2011). Thus, we present networks of this type in the form of case studies. Our discussion is concluded with a short summary.

## 9.2 The concept and ideal-types of corruption

International literature offers numerous definitions for the concept of corruption.[2] This is partly due to the fact that – because of its historical–cultural nature and its versatile character – the phenomenon is a rather difficult one to describe in general terms. The situation is further complicated by continuously ongoing debates between experts in an effort to reach consensus on a general definition (see for example Lambsdorff, 2007, pp. 15–20). The creation and acceptance of a general definition is made difficult by a number of factors. For one thing, the term stands for a number of distinct, but related, phenomena. Moreover, differences become apparent even when examining the issue from the legal and cultural perspectives of a given country,[3] as well as when exploring the common concept of corruption and its effects on public interest. All of this is further differentiated by the fact that

---

[2] The word corruption (Latin: *corrumpo*) originally means to break, to destroy.
[3] Here, there is emphasis not only on the word legal, but also on the word country: as a result of historical, cultural, and other differences, there is considerable variance between corruption phenomena and – following from this – the general concepts of corruption in the individual countries.

the term can comprise the definitions of those with involvement and vested interest in corruption, as well as those who are actively fighting against it (Gardiner, 2009).

In spite of the debated questions, however, of the widely known definitions often referred to by social scientists, Nye's take, certainly warrants mentioning. According to Nye (2008, p. 284), "corruption is behavior which deviates from the formal duties of a public role because of private-regarding (personal, close family, private clique) pecuniary or status gains; or violates rules against the exercise of certain types of private-regarding influence." In contrast, Klitgaard (1991) argues that we can only speak of corruption if the individual places her/his personal interest in an unpermitted manner above the causes (or persons) that s/he is otherwise meant to serve.[4] In the opinion of Rose-Ackerman (1978, 1999, 2006), corruption appears in the simultaneous presence of both wealth and the power of the State. It is markedly characterized by a willingness to employ forbidden (financial) means to influence the decision process. She regards the corruption transaction between a private person (or private company) and the government as its most common form, where the corruptor, in return for a bribe, obtains an unlawful (financial or other) advantage from the corrupted public officer. Similarly to the aforementioned examples, Lambsdorff stresses a number of important attributes in relation to corruption in the definition he utilizes in his research. In accordance with this, corruption – at first take – is "the misuse of public power for private benefit" (Lambsdorff, 2007, p. 1).

These various definitions seem to share some common attributes, which comprise the substantive factors in the innumerable manifestations of the phenomena. These characteristic features – with the inclusion of Lasswell's (1930) points – can be summed up as follows:

- a corrupt transaction can only take place if at least two parties are involved (be it a person, a community, or an institution);
- the premeditated and conscious decision of the parties to participate in the transaction is also a basic condition;
- the intent to exert influence with the aim of obtaining a personal advantage is an integral element of corruption – in other words, there is a pursuit of self-interest, where personal objectives are placed, in the case of political corruption, above public interest, and in the case of economic corruption, above the owner's interest;[5]
- and, finally, breaking the rules (forbidden, illegal, unauthorized activities) is also a significant feature.

---

[4] Klitgaard (1991) presents the various forms of corruption, drawing on examples from different sectors, in reference to the initiators of the corruption.

[5] The meaning of "public interest" has been, and continues to be, a debated question that we do not consider in this chapter.

As Lambsdorff's definition seemed the most productive for our purposes, we will also rely on this in our present analysis. This definition is based on the principal–agent–client model of modern political and institutional economics.[6] The model is grounded in rational choice theory: its actors consider the expected benefits and cost of their options and choose the alternative which promises to yield the highest net benefit. In the model, the agent (e.g. an official who issues construction permits) is entrusted with power by the principal (e.g. local government). During this process, the principal (i) delegates certain tasks to the agent, (ii) determines the formal rules according to which the tasks are to be performed, and (iii) offers remuneration to the agent for completing the tasks, who, in return, (iv) remains loyal to the principal, which means s/he performs the tasks in accordance with the rules that have been laid out. The agent will respond to the client's needs within the specified framework (vi), for example, an application for a construction permit (v). These relationships are illustrated in Figure 9.1 by directed graphs, in accordance with the numbering in the text above.

Corruption takes place when one of these players (in most cases the agent) breaks the rules out of self-interest, thereby hurting the interest of the other players (in most cases the interest of the principal and/or the client). The objective of the principal is to motivate the agent to perform an optimal amount of productive activity and an optimal amount of unproductive – in other words, corrupt – activity. The fundamental problem of the actors originates from the information asymmetry, which is an inherent characteristic of the relationships between those involved.[7] This roughly means that, in the examined scenario, the level of awareness is considerably different for the individual parties. This produces a situation in which those actors who are in possession of private information can opportunistically exploit the information

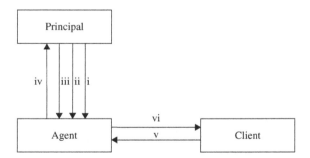

Figure 9.1. The principal–agent–client graph.

---

[6] See Lambsdorff (2007, pp. 62–65). For other applications of the principal–agent–client model to corruption see for example: Klitgaard (1991), Szántó (1999), Andvig and Fjeldstad (2000).

[7] For the concept of information asymmetry see e.g. Rasmusen (1989, pp. 193–226). On the analysis of the negative political effects of informational asymmetry see Szántó (2009).

to their own advantage in an opportunistic manner. The agent, for example, is a lot better informed than the principal about the details of transactions involving the client.

Based on information asymmetry between actors, four ideal-types of corruption can be distinguished in the principal–agent–client model (Lambsdorff, 2007, pp. 18–19), which can be represented using directed graphs. It seems an obvious step to differentiate, within these four pure types, two subtypes of corruption relationships, between the principal and the agent on the one hand, and between the agent and the client on the other – and based on this, distinguish between bribery (A), extortion (B), embezzlement (C), and fraud (D). Klitgaard (1991, p. 50) refers to the former two types as external corruption and to the latter two as internal corruption.

(A) In the case of *bribery*, the client – as the initiator of the corruption transaction – acts as the briber and offers a bribe to the agent. In return, the client procures an advantage in an illicit manner (for example, obtains an unauthorized permit or avoids the disadvantageous consequences of a legal transgression), which s/he could not do otherwise. This type of interaction is shown in Figure 9.2, where illustration of the original relationship is supplemented with graphs representing the bribe and the procurement of an unlawful advantage.

(B) *Extortion* occurs when the agent (extorter) – as the initiator of the corruption transaction – uses his/her power to coerce money (or other benefit) out of the client (extortee). The client must pay for the service (or for speeding the procedure), to which s/he would otherwise be legally entitled. The agent, however, exerts threats, coercion, or even aggression in order to get the client to pay. This scenario is illustrated by Figure 9.3, where the threat and the path of the extorted sum are shown by directed graphs.

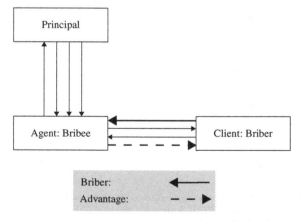

Figure 9.2. (A) Bribery graph.

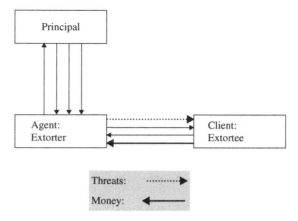

Figure 9.3. (B) Extortion graph.

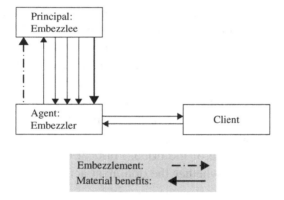

Figure 9.4. (C) Embezzlement graph.

(C) We examine two corruption scenarios within the framework of the principal–agent relationship. One of these is *embezzlement* or *misappropriation*. This is an action initiated by the agent (embezzler), whereby s/he (partially or wholly) appropriates the asset or the right of disposal entrusted to her/his care, and disposes of these as her/his own. In this transaction, the agent (embezzler) inflicts a loss on the principal. Figure 9.4 shows the graph of embezzlement and financial advantages.

(D) The fourth ideal-type of corruption is *fraud*, where the agent, by increasing the information asymmetry, employs hidden action to obtain an advantage. Alternatively, the agent can also actively conceal information from the principal (for example, by forging documents, manipulating information, or other methods). Figure 9.5 shows the graph of manipulating information and obtaining financial advantage.

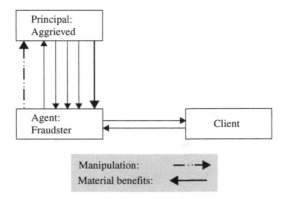

Figure 9.5. (D) Fraud graph.

Thus, these are the ideal-types of corruption that we accept and work with in our research. Naturally, there are numerous other possibilities for classification in relation to the phenomenon. One of these (Lambsdorff, 2007, p. 20) entails an examination of whether it is the briber or the bribee who obtains a greater advantage during the transaction. This is largely dependent on which party has a stronger bargaining position. If the briber realizes a greater advantage, this is referred to as "clientelist" corruption. If, on the other hand, the corruption transaction results in greater advantage for the bribee, we speak of "patrimonial" corruption. Another possibility is to distinguish between petty and grand corruption, where the differentiation is based on the size of the bribe. Political or administrative corruption are differentiated based on whether the dominant actors are politicians or public officers.

Now let us take a look at the possible motivating factors that prompt actors to embed their corruption transactions in personal and institutional networks (Granovetter, 2007). Corruption transactions hide risks. Carrying them through successfully involves various types of costs. While in the case of bribery and extortion, the client and the agent (and/or brokers) bear the risk, in embezzlement and fraud scenarios it is primarily the agent (and/or hidden principal). Monitoring and realizing corruption transactions, obtaining and processing the required information, as well as the bargaining, decisions, and coercion associated with such undertakings carry substantial transaction costs. The more efficient the institutions are in detecting and sanctioning corruption and the more severe the expected punishment if caught, the greater the incurred costs.[8] These two factors also considerably influence the cost of transactions. At the same time, the greater the expected advantage,

---

[8] These correlations in connection with corruption can be articulated based on the traditional microeconomic model of crime (Becker, 1968).

the more likely corruption will occur. As we have every reason to assume that the actors in corruption transactions are capable of gauging the risks and estimating the costs, we can therefore expect that they will attempt to minimize these using all means at their disposal, thereby increasing the net profit produced by the transactions. Among these means, it would seem advisable to consider the establishment, maintenance, and expansion of various types of networks. From this point on, we will differentiate between two basic network forms. We will regard cases where the participants of corruption transactions embed their corrupt dealings in *interpersonal* networks, separately from those where the transactions are embedded in *institutional* (business, contractual, political, etc.) networks.

In the second half of the study, based on the above outlined four ideal-types, we make an attempt to sketch out typical corruption networks. We supplement the simple, three-actor graphs – as previously mentioned – with new actors (hidden principal, broker), hidden roles, and new network contents (personal and institutional relationships). Network contents signify the social and institutional structure – and embeddedness – of corruption transactions (Granovetter, 2007). In our discussion, we first describe a specific case (based on media sources and interview experiences), then we use graphs to sketch out multiplayer, multiplex corruption networks.

## 9.3 A few typical corruption networks in Hungary

As we have seen when outlining the various ideal-types, corruption transactions, in the simplest scenarios, occur between two actors. Corruption transactions between the agent and the client are typically based on a personal relationship of an occasional (or regular) nature. The other subtype of transactions, involving only a few actors, is when an institutional relationship (also) develops between the two parties. In the second typical scenario, the corruption transaction is embedded in a multiplayer network where we can often assume the existence of various personal and institutional relationships between the agent and the client prior to the transaction. It is primarily these relationships that make the corruption transaction possible and pave the way for its repeated occurrence between the same actors. One of the upcoming examples (case number one) shows this situation: the parking inspectors, in a corruption scenario initiated by parking car owners, share in the resulting corruption fee with their superiors. The other type is when the corruptive, or corrupt, agent uses a broker company for both transferring and withdrawing the corruption fee. We will be taking a closer look at such a scenario in the second case study, in which, when a company requests permission from the authorities, it is "advisable" for it to contract the expert recommended by the authorities, for instance, in order to prepare a specific impact assessment.

Table 9.1. *Pure types of corruption transactions involving few actors and many actors, with embeddedness at the interpersonal and institutional level.*

|  | Type of embeddedness | |
| --- | --- | --- |
|  | Personal | Institutional |
| Few | e.g. At the border, a familiar customs officer turns a blind eye to customs offences in return for a bribe. | e.g. A company files for a permit with a state office and is "required" to contract a specified expert for preparing an impact assessment. |
| Number of actors | | |
| Many | e.g. Familiar parking inspectors, in return for a bribe, turn a blind eye to regular customers' failure to pay the parking fee. | e.g. Real estate management by local government, where assets are regularly sold under market value. |

In the light of all of this, it seems wise to draw a line between the various types of corruption transactions, based on the number of actors and the type of embeddedness. In the process, we make our way from simpler transactions with fewer actors and interpersonal embeddedness to more complex affairs, connecting many actors with embeddedness at the institutional level. Two intermediate possibilities between these scenarios are corruption transactions involving a few players at the institutional level and multiple players at the personal level. These pure types are summarized and illustrated through examples in Table 9.1. Personal and institutional embeddedness appear side by side in the majority of real-life situations and, in empirical research, it makes sense to consider them together.

Based on our previous research,[9] it is safe to state that, in Hungary between 2001 and 2009, the ratio of multiplayer, network-based corruption transactions showed an increasing tendency (see Figure 9.6). Reports by online news portals on cases of

[9] In the course of this research, in addition to interviewing company heads, we completed the content analysis of eight Internet news portals for the time period between January 1, 2001 and December 31, 2009 (Szántó *et al.*, 2011). Of the articles available through online sources, we gathered and analyzed in detail those that discussed cases of suspected corruption. We collected cases of corruption in Hungary exclusively. The articles were systematized by arranging news of the same transactions into "corruption cases." We noted the main characteristics of these cases and coded them in accordance with the various features of corruption transactions. Over the examined time period, the topic of corruption comprised the subject of continuous public discourse. Over 3,500 articles have been included in the database and 548 different cases of suspected corruption identified and analyzed.

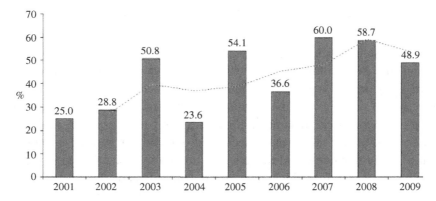

Figure 9.6. Percentage of multiplayer, chain-like corruption cases in the media, 2001–2009 (548 cases of suspected corruption).

suspected corruption indicate that the ratio of transactions suggesting the existence of corruption networks doubled during the examined time period.

These studies make it possible to define typical corruption networks. Since the basic model of corruption gives a good description of two-actor, one-time corruption transactions, on the following pages, primarily multiplayer, chain-like corruption transactions will be analyzed from a perspective of interpersonal and institutional embeddedness. Four cases will be outlined, followed by a graph showing the various types of corruption networks implied by these concrete examples:

(1) Case number one: interpersonal embeddedness of inspection-related corruption;
(2) Case number two: interpersonal and institutional embeddedness of corruption related to the acquisition of permits with broker;
(3) Case number three: interpersonal and institutional embeddedness of hidden principal corruption related to local government real estate purchase;
(4) Case number four: interpersonal and institutional embeddedness of corruption related to the use of EU funds.

### 9.3.1 Bribery and extortion network

*Case number one: personal embeddedness of corruption relating to inspections*[10]

In the center of Budapest, parking fees are charged pro rata. The maximum parking time is three hours. After this, a new parking ticket must be purchased. Alternatively, the parking time can also be extended by making a mobile phone

---

[10] *Source*: articles accessible through the Internet.

call. The parking meters are operated, on the authority of the local governments, by a number of private companies per district, who also collect the parking fees. The collected sum is then shared by the parking company and the local government according to a given ratio. How much is to be paid and which areas should be reserved for paid parking are determined by the local government. Prior to the introduction of parking fees many who conducted business in the inner districts chose to get there by car, thus using up all the free parking spaces. After the new measures took effect, there was a decreased demand for parking spaces, and it was more likely that one could find a space, albeit for a relatively high fee. These regulations work well for vehicle owners who only occasionally need to access the inner districts by car for some odd errand, and manage to take care of their business in less than three hours. The initiators of these regulations, however, did not take into consideration that those entrepreneurs, business and restaurant owners who continue to drive to these areas provide a continuous stream of parking revenue. The employees of the parking company inspect the vehicles and, in the case of exceeded parking time limit or failure to pay the parking fee, impose a fine five times the three-hour parking fee. As paying the parking fee every three hours would require close attention and a considerable loss of time on the part of the entrepreneurs who regularly park in the inner districts, not to mention that a monthly parking pass would be extremely costly, it was worth bribing the parking inspectors, so that they would not check the parking ticket and would not impose a fine. In exchange for this, payments were made to them in cash, weeks, even months, in advance. It sufficed to place a note on the windshield of the car to instruct inspectors as to which shop to pick up the parking fee – or bribe – from. In return, they did not document a failure to pay the parking fee. In time, this type of bribery became such a common and widespread practice that it made sense for parking inspectors to share in the corruption profit with the head of the parking company. Finally, an entire corruption network was created: from the clients to the parking inspectors, to their boss, who shared in the corruption profit. Corruption tariffs were determined for given areas and the head of the parking company regularly demanded a certain sum from the bribes collected by the inspectors, threatening to expose them if they failed to pay.

In this example, corruption developed through personal relationships:[11] since the driver of the vehicle frequently parked in the same area, which constituted the inspection zone of a few parking inspectors, personal relationships unavoidably

---

[11] In this example, corruption was brought about by the incorrect pricing of paid parking. If the rule makers had differentiated between the two groups of consumers – regular parkers and irregular parkers – and made possible the purchasing of a yearly parking pass with some sort of discount for the latter, a situation conducive to corruption would have had less (or no) chance of developing.

developed between the parties during the inspection and fining process. Consequently, corruption resulted from a kind of individual "bargaining." Those involved found a mutually beneficial solution to avoid payment of fines. In our interpretation, this is bribery: the transaction was initiated by the client, or the driver of the parking vehicle. Through personal relationships, the corruption transaction became a general and predictable practice. During repeated transactions, the inspectors recognized the given entrepreneur's vehicle, for which, in the absence of a valid parking ticket, they did not impose a fine. Instead, once a week (or month), they sought out the driver for the bribe. A different type of personal relationship came into play when the head of the parking company became aware of this phenomenon. After a while, the company head also demanded a share of the bribe, in return for not exposing the already existing corruption network. In this scenario, the original principal (the company head) abandons his/her primary role for that of the hidden agent, while the inspectors suddenly find themselves in the role of the hidden client, as their boss, in the role of the hidden agent, applies extortion by coercing them to provide to him a share of the bribe.[12]

Figure 9.7 outlines the above situation with the help of graphs. It simultaneously illustrates the bribery between the parking inspectors and the vehicle drivers, and the extortion between the company head and the parking inspectors. The presence of a personal relationship between the agents and the clients, as well as between

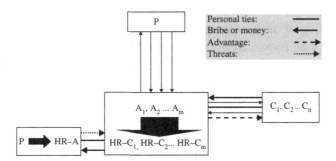

Figure 9.7. Graph of networked bribery and extortion. Notation: $P$ = principal (head of parking company); $A_1, A_2, \ldots, A_m$ = agents (parking inspectors); $C_1, C_2, \ldots, C_n$ = clients (vehicle drivers frequently parking in the same place); the box at the bottom left illustrates the principal switches roles = practices extortion upon the agents in a hidden agent role ($HR - A$: hidden role – agent); the largest box shows the agents switch roles = in a hidden client role, in response to the extortion, they pay a share of the bribe ($HR - C$: hidden role – client).

---

[12] The entire transaction was uncovered because an extorted inspector was unwilling to cooperate with his superior. He refused to hand over a certain amount of the corruption money, and was consequently fired. Afterwards, he reported the incident to the police and made a disclosing statement exposing the entire operation.

actors that transform from principal to hidden agent and from agent to hidden client, is an important condition for the development of this type of corruption network.[13]

### 9.3.2 Extortion network

*Case number two: personal and institutional embeddedness of corruption related to obtaining permits with broker*[14]

In Hungary, the construction of wind power plants requires 15 to 30 different permits and/or licenses. This number changes depending on the number of partner authorities whose collaboration is required by the environmental protection authority in order to issue a so-called environmental license. Investors need to obtain this license in order to submit a request to the Hungarian Energy Office in response to a tender for expanding wind power plant capacities (Tóth, 2010). The time required to obtain this license can be one to two years, depending on the amount of money and energy the investor is willing to devote to expediting the process. The path of the submitted requests can be tracked between the various offices and expedited through a series of corruption transactions. Our interviews have shown that, in this type of corruption transaction, the license issuing body is usually the initiator, which makes this extortion. The submitted request arrives on the desk of a representative of the issuing authority. When the investor enquires about the status of the case, the agent initiates a meeting with the client – a representative of the investor – to take place in the office. The objective of this meeting is to clarify the details of the proposed project, the formal and contentual errors in the license application, and the scheduling of the necessary modifications. During the discussion, the official suggests that an impact assessment should be conducted in order to obtain the license. The objective of the assessment would be to examine the effects of the project in question on the immediate natural environment – the surrounding animal and plant life. The investor is also informed that, in the interests of expediting the license acquisition process, a certain Company X should be hired to prepare the assessment. Our interviewees gave accounts of various impact assessment topics: the effects of the wind power plant on flying invertebrates (e.g. flies), the effects of the wind power plant on the migration paths of tree frogs, the effects of the wind power plant on the surrounding forest biota, etc. The client is put in a difficult situation: s/he would not like to miss the Hungarian Energy Office's application deadline, so it would be advisable to expedite the license acquisition process. The recommended company is thus commissioned to conduct the impact assessment

---

[13] A more complex corruption network can develop if there are additional employees in charge of monitoring the activities of the parking inspectors, thus wedged between the latter and the head of the company. Similarly to – or in cooperation with – the company head, they can also extort the corrupt parking inspectors, threatening to expose their practices.

[14] *Source*: interviews conducted by the authors.

study. General experience shows that once the completed impact assessment is attached to the license request, approval follows within a few days, enabling the investor to move on to the next phase of the process.

In the above mentioned corruption scenario, Company X appears between the agent and the client as the agent's broker, who facilitates concealment of the corruption transaction in two different ways. On the one hand, no direct financial transaction takes place between the agent and the client, on the other, the payment of the corruption fee (in the form of payment for the completed impact assessment) and provision of the corruption gain take place at different times. It is possible that the agent only receives the corruption fee long after provision of the corruption gain (issuing the license). While in the case of personal affiliation, this means a payment into the pocket, in an institutional relationship, this transfer takes place based on a contract signed between the agent and the broker. These two factors serve to conceal the corruption transaction and considerably decrease the risk of getting caught. All of this is made possible by the utilization of a broker. The defining mechanism here is the presence of a pre-existent relationship between the agent and the broker, which precedes the establishment of a connection with the client. This relationship is usually personal, but it can also become institutionalized: for instance, the agent may have partial ownership of the company that conducts the impact assessment. If, however, a personal connection exists between the agent and the broker, the broker can be a close friend or acquaintance of the agent. It is also possible that the agent has a relative as partial owner of Company X. Thus, this type of corruption is embedded in personal and institutional relationships. In the absence of these, the corruption transaction cannot take place. The personal or institutional relationships which exist between the agent and the broker can, of course, be established independently from the corruption, for other reasons. It is also possible, however, that they were formed in order to facilitate extortion transactions. In this case, the agent (or a family member, acquaintance, etc.) may set up a company expressly to receive the corruption fee. This, in turn, can result in a solidification of the corruption network, as embedded in institutional and/or personal relationships. The agent can also extort a corruption fee from the next client in a similar manner, with a low risk of getting caught. In the course of this activity, the transaction costs that arise with the establishment of the network are recovered. The corruption network that is sketched out by this specific example (Figure 9.8) can, according to what we have gathered from our interviews, be considered as typical, often coming into existence in other areas of economic life in connection with the acquisition of other types of permits and licenses. Thus, in this situation, the network of connections, in comparison to the ideal-typical scenario of extortion, extends to a broker, whilst also becoming embedded in personal and institutional relationships. Another possible interpretation of this situation is that the role of the

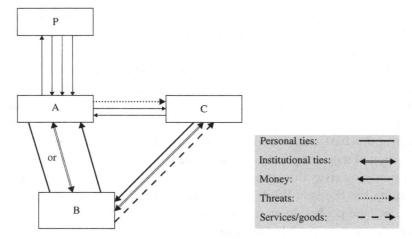

Figure 9.8. Graph of networked extortion. Notation: *P* = principal (license/permit issuing body); *A* = agent (official of license/permit issuing body); *C* = client (investor applying for a license/permit); *B* = broker (Company X hired to prepare an impact assessment report).

broker is aimed at bridging the "structural hole" of corruption between the agent and the client (Burt, 2005).

### 9.3.3 Embezzlement network

*Case number three: personal and institutional embeddedness of hidden principal corruption in connection with real estate purchase by the local government*[15]

An internationally owned company group in Hungary wanted to sell one of its office buildings, but they could not find a buyer. Someone at the company came up with the idea that the local government may be interested. A representative from the company contacted the local government and offered to sell them the aforementioned property for a given amount of money. The leader of the financial committee of the local government responded: "Why not sell it for 10% more? Then we will buy it." The company representative did not understand the question at first. Later, it became clear to him that he was to get 10% of the increased purchase price to the head of the financial committee, in cash. It also became apparent later that this was no simple corruption transaction: the head of the financial committee was backed by the treasurers of the two (competing) parties, which constituted the majority of the municipality, who had agreed that, if the seller brought the 10% to the local government office in cash, the body of representatives of the local government would vote in favor of signing the purchase agreement. And this is exactly what

[15] *Source*: interviews conducted by the authors.

happened. The company representative withdrew the money from a foreign account, opened expressly for this purpose, and delivered the agreed upon sum to the local government office at the specified time. The financial representatives of the two parties counted the money, after which the body of representatives approved the purchase. In order to complete the payment transaction, the company had to have funds set aside for the so-called "below the line" expenses. In reference to this, the company representative explained: "corruption requires the existence of a certain infrastructure."

In the above case, the roles of the principal, agent, and client can be illustrated by introducing – next to the principal (the person in charge at the local government: the mayor), the agent (the person in charge of the financial committee by designation of a party), and the client (the representative of the company selling the real estate) – the treasurers of the two political parties, or the hidden principals, into the equation (see Figure 9.9). Although, here, the corruption appears to consist of a two-actor, one-time transaction, it is not. Firstly, the transfer of the corruption fee in cash already presumes that the corruptive company has the required background: a foreign bank account is opened in order to make the payment of the corruption money. Secondly, it is not really the number one leader of the local government that stands behind the agent as principal, but the hidden principals, in the person of the political party representatives. Thus, the transaction presupposes unique relationships on both sides. In the case of the client, in response to the proposition of corruption, "below

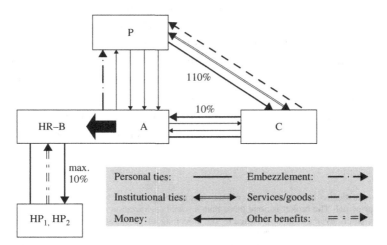

Figure 9.9. Graph of networked embezzlement. Notation: $P$ = principal (mayor); $A$ = agent (head of the financial committee of the local government); $C$ = client (real estate seller); $A - HR - B$ = the agent switches roles: in the role of a hidden broker gets max. 10% to the "hidden principals" ($HR - B$: hidden role – broker); $HP_1, HP_2$ = hidden principals (treasurers of the political parties).

the line" funds are set aside. In the case of the agent, belonging to a political party preceded the transaction. In fact, it was probably this very circumstance which prompted the corruption transaction.[16] As, in the examined example, the client sells the real estate to the agent for 10% more; thus an institutional connection is established. The client, however, gets this 10% difference in the purchasing price through her/his personal connections to the agent, who, in the meantime, has switched roles and passes the money on to the hidden principals (presumably in return for a brokerage commission or other advantage), as the hidden agent of the parties. All of this comprises the complex pattern of embezzlement, as generated by party financing.[17]

### 9.3.4 Fraud network

*Case number four: personal and institutional embeddedness of corruption in connection with the use of EU funds*[18]

The ABC institution has won a grant of tens of millions of Euros for a project aimed at analyzing and monitoring different labor market processes. The system had to be set up in two years and, during this time, the institution had to spend the grant. This objective was accomplished by the leaders of the institution by purchasing and paying for studies which had already existed prior to the start of the project, but were only known in small professional circles. In addition, they employed hundreds of external experts, without them doing any actual work, and bought services (e.g. research) that were not closely related to the original objectives of the project. The resulting background studies and analyses were mostly useless in terms of realizing the original goals. Nor did the project materialize in accordance with the initial aims. While the realization process could hardly be considered successful from a professional standpoint, from a financial perspective everything progressed according to regulations. Partners, when needed, were selected through public procurement, contracts were signed, the performance was documented and performance certificates were issued prior to payment.

---

[16] In a more sophisticated model, the agent already has a broker through whom the money makes its way to the hidden principal. The broker is none other than a social organization or foundation with close ties to the hidden principal, or even a company owned by the hidden agent. In this scenario, the client is coerced by the principal to sell a service to the broker for less than the market rate. The corruption fee – profit from purchasing the service at a submarket price – is then collected by the hidden principal. This model presupposes a more complex network of institutional relationships. The broker, who plays a key role in minimizing the risk of getting caught for the transaction, is of equal importance to both the agent and the hidden principal.

[17] Embezzlement often manifests in conjunction with bribery. This is also evidenced by the fact that, during criminal proceedings of this type, from a standpoint of investigation strategy, these two cases are usually handled together (Ibolya, 2010).

[18] *Source*: interviews conducted by the authors.

When it comes to the use of EU grants, it is in the interest of the agent (project implementer) from the beginning to exaggerate the budget as much as possible, in other words, to make as many investments and buy as many services – which are only loosely related to the initial objectives – as possible. This interest may perversely coincide with the interest of the authorities that make decisions about, and monitor the project, to not waste any funds granted by the EU. Rather, the objective is to "fill them with the appropriate professional content," thereby increasing the country's "absorption capacity." For these authorities, it is easier – or it carries smaller specific transaction costs – to organize and manage a large-scale project than many smaller ones which, in total, require the same amount of funds as the latter. The agent then launches the project with his or her attention primarily focused on the scheduled spending of the allocated amount, with realization of the original goals being only of secondary importance. The controlling authority chiefly monitors the observation of deadlines and scheduled spending. The effective use of funds demands that the agent (project implementer) has the appropriate personal and institutional relationships, in order to gain access to those quasi-performances that, to some extent, fit the announced objective, make the spending of the project fund and, thus, the "successful" completion of the project possible. During the implementation process, it is not necessary to resort to bribery, but the system of mutual aid – logrolling – is a widespread phenomenon, and this is what facilitates the "successful" completion of the project. The experts and institutions (clients) selected by the agent aid the project implementer (the agent) in finding work and other sources of funding in other areas, for example, in other EU projects. To the controlling authority (or even to the EU), everything seems to be in order: the grant was spent, the project was realized, the expenses were conducted, and the financial report was completed by the book – except that the original goal was not realized. In this respect, the act of fraud manifests in the relationship between the EU (and the taxpayers of the EU) and the project-launching and controlling authority, on the one hand, and between this authority and the implementing institution on the other. The agent (the implementer of the project), in this context, can also act in accordance with the interests of a hidden principal, when transferring a part of the handled EU funds to a colleague, an expert, or a subordinate organization, in return for some quasi-performance. The hidden principal can be a private company, a political party, or a state institution (see Figure 9.10). The establishment of this kind of fraud network is made possible by personal connections between the agent and the hidden principal, the relationship between the agent and the clients, who act as the hidden broker, as well as the institutional (contractual) affiliation between the latter and the hidden principal. As a unique feature of the relationship between the agent and the brokers, it is in this context that the contractual arrangements are made for performances that are more or less irrelevant from the standpoint of

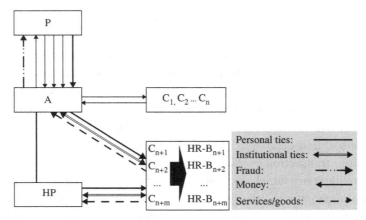

Figure 9.10. Graph of networked fraud. Notation: $P$: principal (authority announcing the call for projects); $A$: agent (implementer of the project); $C_1, C_2, \ldots, C_n$: clients (experts and institutions realizing the original objectives of the project); $C_{n+1}, C_{n+2}, C_{n+m}$: quasi-clients (clients involved in the fraud); $HR - B_{n+1}, HR - B_{n+2}, HR - B_{n+m}$: the quasi-clients switch roles and cater to the needs of the hidden principal in the role of the hidden agent ($HR - B$: hidden role – broker); $HP$: hidden principal (e.g. private company, political party, or state institution).

the original project goals, whereby brokers are significantly overpaid in return for quasi-performances. At the same time, the brokers, during the process of fulfilling the conditions of the contractual agreement with the hidden principal, are satisfied with a smaller compensation. Thus, in the case of hidden party financing, for instance, remuneration for the kind of expert work that is important to the party, can come from the EU project.

## 9.4 Conclusion

We have examined the phenomenon of corruption by focusing on the relationship networks between the actors. In our analysis of corruption, we took the principal–agent–client model as our point of departure, and it was also based on this that the subtypes of corruption were identified: bribery, extortion, embezzlement, and fraud. Specific cases were discussed to supplement the models with the various types of relationships between the actors, as well as the new actors of corruption networks: the brokers and the hidden principals. Moreover, in a number of cases, we also took into account the hidden roles of those involved in the transaction. By this we meant that, as the corruption network develops, some of the actors abandon their original roles and assume new ones (hidden agent, hidden client, and hidden broker). We differentiated between two basic kinds of relationships: personal and institutional. On the other hand, we also made a distinction between two different subtypes based on the number of actors: few-actor and multi-actor relationships.

We demonstrated four possible corruption networks, based on data obtained from interviews and Internet news sources. These scenarios were analyzed according to the four basic corruption types. During this process, we took into account the interpersonal and institutional embeddedness of the different types of corruption networks, illustrating these in multiplex graphs. Based on our research experience, the formation and embedding of these typical network configurations can be expected to occur primarily due to a decrease in the transaction costs and risks associated with carrying through corruption transactions.

In the light of all of this, one of the important conclusions we can draw from our research is that, while corruption transactions can be traced back to a few well-defined and basic types, the presented cases, which can be considered typical, demonstrate the complexity of the manifesting corruption networks and the multiplicity of relationships between the involved players. In a more complicated corruption transaction, for example, both the agent and the client can have a broker; in other words, the payment of the bribe does not take place between the actual actors, but through the brokers. In some cases, the agents, following their own interests, may initiate extortion or fraud, or may accept the bribe offered by the client. In other cases, however, the hidden relationship network is shaped by the interests of the hidden principals, in which personal and institutional intertwinings also play a role.

What could explain the multiplicity and complexity as well as the growing number of players and roles, as demonstrated by these examples? In answering this question, we stress the influence of two factors:

(1) Every corruption transaction comes into being when the given regulatory and institutional conditions are present. These regulatory and institutional frameworks – which already exist prior to the transaction – fundamentally influence how the concrete transaction manifests. In other words, it is the institutional embeddedness and regulatory environment of corruption that determines the framework of its realization. Brokers cannot be utilized, nor can the bribe be "transferred," in bribery during a roadside check. It follows from the situation that this is a simple (didactic) relationship, where the bribe is slipped into the police officer's pocket. If a business venture applies for a license, brokers can enter into the transaction, so that it can also be realized through institutional relationships (e.g. selling and purchasing services or goods between ventures).

(2) On the other hand, the realization of corruption is also influenced by the actors' calculations with reference to the transaction: how big is the risk of getting caught, what is the punishment if that happens, and what is the expected benefit

(or profit) if the corruption is realized? In this regard, we can consider the formation of the various types of corruption networks as the result of these calculations. Multiplayer corruption scenarios that are grounded in institutional relationships carry higher transaction costs than simple setups involving only a few parties. At the same time, through the first mentioned type of transaction, the risk of getting caught can be reduced for all involved. Through the utilization of brokers, for example, collecting the bribe can separate in time from provision of the service, and, in fact, by availing themselves of the broker(s)'s services, the original actors can conceal their corrupt transactions as legitimate dealings.

# Part III

Crisis, extinction, world system change:
network dynamics on a large scale

# 10

# How creative elements help the recovery of networks after crisis: lessons from biology[1]

ÁGOSTON MIHALIK, AMBRUS S. KAPOSI, ISTVÁN A. KOVÁCS,
TIBOR NÁNÁSI, ROBIN PALOTAI, ÁDÁM RÁK,
MÁTÉ S. SZALAY-BEKO, AND PETER CSERMELY

In the middle of a world economic crisis, studies of biological crisis-survival strategies, which have proven their efficiency over a billion years of evolution, may provide novel ideas and solutions. In this chapter, we summarize our knowledge and the latest results which show that the "crisis of cells," called stress in biology, and its extreme form, programmed cell death (also known as apoptosis), induce a decrease in the inter-modular contacts of protein–protein interaction networks of yeast and human cells, respectively. Moreover, programmed cell death leads to the decomposition of several key modules of human cells. This network disintegration fits well with the general scheme of network topological phase transitions upon a decrease of resources (and/or an increase of perturbations, stress). The modular disassembly resembles changes to social networks in crisis-like situations. Thus, the re-association of cellular networks after stress and the necessity of mobile, inter-modular elements ("creative elements") for this process of "modular evolution" may give us adaptable schemes for crisis-survival strategies, proving the old observation that a crisis is not a disaster but a challenge providing a chance for development.

## 10.1 How can biological networks help our coping with crisis situations and our understanding of social networks?

From the second half of the twentieth century, the increasing complexity and the information-richness of our everyday life have got closer and closer to the limits of the information handling capacity of our logical, left-hemisphere brain. The Western civilization has been increasingly using (and actually, over-using) logical thinking

[1] Work in the authors' laboratory was supported by the EU (TÁMOP-4.2.2/B-10/1-2010-0013) and the Hungarian National Science Foundation (OTKA K83314).

from the very beginning and, especially, from the Enlightenment of the eighteenth century, signaling the birth of modern scientific thinking. Our schools socialize us and our children to accommodate a conceptual framework, which makes the logical assessment of the changing situations a Pavlovian conditioning of modern times. This way of human cognition is efficient, useful, and has enabled a lot of progress in the past centuries. However, the plethora and complexity of information we "enjoy" today easily saturates the left hemisphere of our brain, which is responsible for logical thinking. Our brain usually has only a few cognitive dimensions (five to seven in males and a slightly higher number in females; Dunbar, 2005), meaning that our left hemisphere can tackle only a few independent ideas simultaneously. The current information load easily surpasses this amount. The situation becomes even worse in times of crisis, since in crisis the generation speed of novel information, the need to get more and more information, and the number of seemingly important pieces of information all increase, which magnifies the perceived information load. We may sum all of this up in a rather extreme form: in crisis the logical brain gives up and emotions, instincts, and panic-like herding behavior become dominant.

How can networks help to cope with panic-like behavior in crisis-prone situations? Besides the logical analysis of the left brain hemisphere, the subconscious summary of billions of previous experiences in the form of instincts or emotions, and giving up any analysis, and simply copying what others do, herein we have another way of information perception, the holistic, pattern-like recognition of the right brain hemisphere. With this ability we "immediately" recognize the focal points of an incomprehensibly large dataset – as though it were converted to a picture-like form.

Networks became an increasingly popular method of analysis in recent decades, "conquering" scientific disciplines one after the other from the starting point, sociology. One of the major reasons for this network-revolution is that networks provide a dual way of cognition: they both allow the assessment of the dataset from an image (the visual image of the network), and provide an easy and straightforward path to "number-crunching" of the most important segments. In the visual images of networks we can easily recognize their

- hubs (Barabási and Albert, 1999);
- communities (Girvan and Newman, 2002; Kovács *et al.*, 2010; Palla *et al.*, 2005);
- bridges connecting distant elements and distant communities (Watts and Strogatz, 1998; Csermely, 2009);
- the highly versatile, inter-modular "creative elements" (Csermely, 2008),

and many other key features. After the identification of these points, networks provide a fast way to identify the exact data, such as the identity of the elements, the strength of their links, etc. Most of the above, key structural segments can

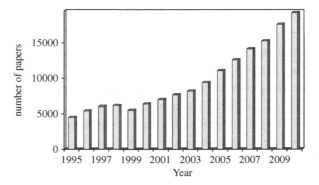

Figure 10.1. Propagation of the network method. The number of publications containing the words "network" or "networks" in their title or abstract was collected from the database www.pubmed.com. The 2010 data is an extrapolation. Notice that the two seminal papers on networks were published in the late nineties (Watts and Strogatz, 1998 and Barabàsi and Albert, 1999). We should highlight that the quality of data are somewhat compromised since the word "network" may also refer to scientific collaboration networks, and the total number of articles per year in PubMed also increased over this period.

also already be assigned using precise algorithms. With these dual possibilities, networks help to mobilize both hemispheres of our brains to cope with the increased information-load in crisis.

The search for the "essence" is as old as human culture itself. Why would networks provide a better solution for this evergreen problem than the myriads of other approaches? Their visual image alone would not explain the tremendous success networks enjoyed and enjoy in the last decade (Figure 10.1). The visualization of the hundred-million users of a mobile-phone network would make a rather incomprehensible picture, even with the hierarchical zooming-in techniques of most modern visualization options (Kovács *et al.*, 2010; Pavlopoulos *et al.*, 2008). In past decades, one of the most important and most surprising conclusions of network studies was that several key network features are highly similar in many networks which describe entirely different complex systems. Thus, networks of macromolecules, cells, organisms, ecosystems, and social networks, are similar

- in their small-worldness (Watts and Strogatz, 1998);
- in their close-to scale-free degree distribution (Barabási and Albert, 1999);
- in the hierarchy and nestedness of their layers of increasing complexity (Ravasz *et al.*, 2002);
- in the stabilizing role of their weak links (Csermely, 2009), and in many other features.

This similarity allows for very exciting play: if we find a novel feature of a network of a certain complex system, we may try to find a similar feature of a network of an entirely different complex system. This not only gives a broader scope to our findings, but also allows us to grade their importance. Highly specific network features may not claim general applicability. Additionally, these trans-disciplinary, network-aided studies also help our general understanding. Human cognition is many times hindered by the cognitive barriers of our words. The meaning of words requires a certain context. This gives a rich texture of meaning, but also imprisons the meaning giving it limited scope. Let me give an example. If I say, "aging" every one of us immediately imagines something, an old grandpa, wrinkles, whatever. This immediate recognition of aging as a human feature makes us uneasy about thinking about questions like: "Does the Internet age?"; "Is the structure of our society aged?"; "Can the ecosystem of the whole Earth, Gaia age?" Network science may give us generally applicable tests for phenomena like aging, which make such questions testable. In our chapter we will sum up our knowledge and our own results, giving another example, where a network of a certain layer of complexity (i.e. the protein–protein interaction network of our cells), can teach us patterns (i.e. in this case cellular crisis-recovery patterns), which can be applied to an entirely different situation, the recovery from the current economic and social crisis.

## 10.2 Crisis responses of networks

At the time of writing this chapter, world-economy gurus had started to convince each other that we had slowly emerged from the 2008 monetary and economic crisis. We will start to get those datasets whereby the 2008 crisis and recovery behavior of human societies can be studied only some years later. However, we need answers now. Regrettably, during previous crisis events data collection and analysis were not at the point that they are today, and we have only very few reliable network-like datasets, useful for studying social crisis events of the past. What can we do in such a situation? We may use the analogies described in the introduction to our chapter and examine crises in other networks in the hope that the features we find will be applicable to social networks.

The crisis in cells may provide a good example of this cross-disciplinary approach. When does a cell suffer a crisis? The easy answer is: "always". An event of cellular crisis is called stress. We can induce cellular stress rather easily. We just have to heat or chill the cell, provide too much or too little salt, sugar, oxygen, etc., all of these will lead to a cellular crisis, if we apply them abruptly enough. Our own studies examined the changes in yeast and human protein–protein interaction networks under various forms of stress.

In protein–protein interaction networks, elements are cellular proteins and their links are physical interactions between them. Regrettably, the yeast protein–protein interaction network (interactome; Stark *et al.*, 2011) has not yet been measured after stress. Therefore, we had to choose an indirect method: we modeled the changes in weight of protein–protein interaction links as changes in the abundance of the two messenger ribonucleic acids (mRNA-s) coding the two constituting proteins in stress (Gasch *et al.*, 2000; Holstege *et al.*, 1998). The multitude of our control examinations (Palotai *et al.*, 2008; Mihalik and Csermely, 2011) and other studies (Halbeisen and Gerber, 2009) show that this assumption holds fairly true. We analyzed the overlapping modules of the yeast interactome using our ModuLand method (Kovács *et al.*, 2010). In the first phase of work we had dozens of highly lucrative hypotheses. However, after the application of various controls (different stress types, starting conditions, interactomes, averaging and weight calculating methods, etc.) all but one of our ideas were proved to be false. The only finding, which survived all controls was that the overlap of yeast interactome modules significantly decreases in many types of stress, including heat shock, osmotic shock, and oxidative stress.

Does a decrease in modular overlaps have any importance? Modules of yeast protein–protein interaction network correspond nicely to various key functions of the yeast cell, such as protein synthesis, protein degradation, regulation of the cell cycle, and others (Palotai *et al.*, 2008; Kovács *et al.*, 2010). A decrease in the overlap means that we have many fewer proteins in stress, participating in more than one cellular function. This helps cellular adaptation to stress in the following ways:

- A decrease in inter-modular links is an important factor in sparing system resources, since the maintenance of all contacts requires additional resources from the cell. Energy is a very high-cost item during cellular crisis, when mitochondria (the cellular power stations) become damaged.
- A decrease in overlap between network communities enables more independent functioning. This increases the adaptation potential of the cell, since various segments may explore a larger adaptation space, if they are no longer bound so tightly to each other.
- The decrease in overlap prevents the easy spread of damage. If a cellular module becomes damaged (e.g. one of the proteins is oxidized by the free radicals coming from aberrant mitochondria, the free radicals emanating from the oxidation of this protein will not damage its previous partner in the adjacent module). Similarly, the increased noise coming from the malfunctioning of damaged proteins, membranes, and other cellular components will not disturb all of the cellular communities, since it is arrested at the modular boundaries (Szalay *et al.*,

2007). This latter feature makes isolation of modules similar to the well-known quarantine applied to prevent the spread of infectious maladies.

If cellular crisis is extremely severe, cells often go along a programmed cell pathway called apoptosis (Sőti *et al.*, 2003). This death is altruistic, since apoptosis makes a "neat package" from the cell, preventing the spread of its internal content to the neighborhood. The alternative pathway, cellular necrosis, leads to a break in the cell wall, and to widespread inflammation due to the leak of the internal cellular content.

In programmed cell death, a family of protease enzymes, called caspases, cleaves several hundreds of cellular proteins. Many of these proteins are essential for cellular life, therefore one may wonder why we need cleavage of so many of them. In our studies, we examined the changes in protein–protein interaction networks of human cells after the removal of these caspase-substrate proteins, cleaved during the process of programmed cell death. To make a long story short, we concluded that the apoptosis of human cells separates the modules of the human interactome rather similarly to the modular disassembly we have observed in yeast stress. However, as an additional feature, caspases also induced the disassembly of key modules, leading to a much more complete disintegration of the whole network than in the case of "simple" stress in yeast. Moreover, caspases were extremely efficient in finding the most important human proteins to break the network structure of the human interactome.

## 10.3 Network rearrangements after crisis

So far we have shown two examples, where protein–protein interaction networks are disassembled in a moderate way (stress), or in a more complete fashion (programmed cell death). What happens after the initial, deteriorating phase of crisis? Obviously, the cell must then be re-assembled. Understandably, this task is impossible after the completion of programmed cell death. However, re-assembly of the yeast interactome after stress is a common, straightforward process (Figure 10.2).

How does the cell "know" how the disassembled modules should re-assemble? Obviously, not all inter-modular contacts are destroyed during moderate stress, such as those examined in our studies. Therefore, the remaining contacts guide re-assembly of modules. This ensures that the "new," re-assembled cell will behave in a "cell-like" way, i.e. similarly to the cell before stress.

However, besides the above mechanism of conservation of cellular "rules, norms, and morale" re-assembly has another element too, which inflicts change. The cell will re-assemble in a slightly different structure from that before the stress. These changes may be called an adaptation, or learning process of the cell. They provide

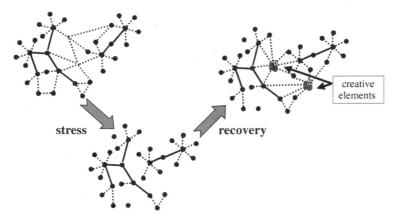

Figure 10.2. Changes in network structure during and after crisis. The illustrative figure shows the behavior, of the yeast interactome during and after a stressful event. During stress, network communities become disassembled, while after stress they re-assemble again – in a slightly different manner. Solid and dashed lines represent strong and weak links, respectively.

a "stress memory" or "stress history" of the given cell, which makes it unique amongst all other cells with identical genetic content (among many other reasons, this is also why two cloned humans may not be entirely identical after the first stressful event their zygotes survived). In the long run the "stress history" of cellular genealogies contributes to the gene-independent, epigenetic mechanisms of evolutionary change, called "modular evolution" (Korcsmáros *et al.*, 2007; Szalay *et al.*, 2007).

What type of behavior of network elements is necessary for changes in the reassembly process after stress? The ability to build "quasi-random," low affinity contacts (weak links) plays a prominent role in this process. If an element has a large variety of weak links, and these links point to distant segments of the network, this element has the ability to change one or a few of its contacts, thus having a large adaptation potential in stress or crisis. Naturally, if this element has contacts to distant network segments, than it is in the overlap of multiple modules. Having this central position it may have a decisive role in building up the novel elements of inter-modular contacts after stress. The features of such elements have been summarized and are called "creative elements" (Csermely, 2008), since they behave similarly to creative people requiring the presence of many independent views and ideas so as not to be bored to death. Excitingly, these creative elements have a major role in introducing innovations to networks from protein structures, where they form the active centers of enzymes, to social networks and ecosystems (Csermely, 2008).

In cells, creative elements are well exemplified by stress proteins. These proteins are bona fide helpers for the survival of cells during stress. Their pivotal role in evolutionary changes had been uncovered a decade ago (Rutherford and Lindquist, 1998). In the past 4–5 years it has become increasingly clear that the scope of creative elements of protein–protein interaction networks goes much beyond stress proteins (Csermely, 2009). As a recent example the yeast genome, which contains approximately 6,000 genes in total, has roughly 300 creative elements (Levy and Siegal, 2008).

### 10.4  How can we adapt the crisis responses of biological networks to social networks and to our own behavior?

The crisis responses we have shown so far are not specific to protein–protein inter-action networks. Another type of cellular networks, mitochondrial networks, also first assemble, and then disassemble later, as the stress increases (Tondera *et al.*, 2009). The crisis-induced disintegration of networks can be observed in various types of inter-firm contacts (Saavedra *et al.*, 2008; Stark and Vedres, 2006). In fact, the essence of such transitions is reminiscent of the Schumpeterian term, "creative destruction" (Schumpeter, 1942).

The generality of network behavior after a decrease in network resources has already been suggested by the work of Tamás Vicsek and co-workers (Derényi *et al.*, 2004). The model shows that as resources available for the formation and maintenance of new contacts decrease (the amount of disassembly-inducing stress increases), the network topology changes from a random graph structure through a scale-free topology and a star network to a complete disassembly to subgraphs (Figure 10.3). These topological phase transitions can be generalized and a number of additional examples can be given, when similar topological changes occur in cellular networks, and in animal communities (Csermely, 2009). The crisis-induced changes of protein–protein interaction networks correspond well to the "scale-free to star" network transition in the case of yeast stress (marked by arrow 1 on Figure 10.3) and to the "star network to subgraphs" transition in the case of the human programmed cell death (marked by arrow 2 on Figure 10.3), respectively.

The changes we have described encourage us to think about the possible analo-gies of crisis-induced changes in social networks. Our common knowledge is in agreement with the possible consequences. The story of our cells is our own story too. In the first shock of crisis, we tend to become more isolated, and this relative isolation is true all the more to the contacts leading to distant social groups. The society breaks into more isolated "islands," where the members of groups deal with their own problems. The isolation of social groups increases their tensions and possible conflicts.

How can we reassemble the social network after crisis? The first step is to enforce the islands of trust. Such islands can be very stable, even in a completely different

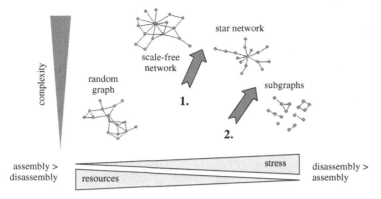

Figure 10.3. Changes of network structures upon increasing crisis (stress). The figure shows the topological phase transition of networks over a larger range than that shown in Figure 10.2. The "stress-scale" of the figure spans from the no-stress situation (left side, where assembly of contacts dominates over disassembly), to the completely severe, "lethal" stress (right side, where disassembly of contacts dominates over assembly). The vertical position reflects the approximate complexity of the network structure. Arrows 1 and 2 point to the approximate position of the yeast stress and the human programmed cell death conditions of our experiments, respectively.

environment (Wang *et al.*, 2008). However, this first step is not enough. From our own island of trust we have to be adventurous and discover the neighboring islands. For this we have to increase the number of social dimensions we have. With a rich structure of social dimensions we may reach a number of novel groups, with novel information, novel abilities, and solutions to the crisis. We may increase our value to be mediators and brokers for all of these conflicts and misunderstandings, which have been sadly enriched during the crisis. With each misunderstanding solved, we decrease the possibility of a future conflict. The novel contacts and the richness of social dimensions help us a lot to transform the conflicts to a new frame, where the original, adamant opposition can be seen, and ridiculed from a new distance. The wider scope of contacts also enables us to "think big," and instead of competing, cooperate to explore novel options together.

Such behavior needs a stable and positive self-esteem. What can we suggest to those who do not yet have this? They should dare to make the very first step. They should leave those who are similarly desperate to themselves and they should make contacts with those who are already able to build trustful contacts. Give a positive gesture, give a smile. And a virtuous cycle begins: they get a positive feedback which enforces their positive self-esteem. And a more positive self-esteem helps the next smile. This "smile-propagation" is illustrated by the examination of Web-communities, where smiling faces are mostly surrounded by other smiling faces, while sad figures draw other sad companions (Christakis and Fowler, 2009).

This summary has shown that the key to crisis recovery is in our own hands. Crisis is not a disaster, but a chance to develop something new, a chance to change. However, the key to crisis recovery is a funny one. None of us has it alone – but we all have it together. It is time to look around.

# 11

# Networks and globalization policies

DOUGLAS R. WHITE

This chapter argues for connecting models of several kinds of macro- and micro-processes as they affect structure and dynamics in the globalization of networks of trade. The purpose is to explore multiple levels of structure, process, and adaptation and to loosen assumptions about determinacy in models of networks and globalization. As do many models of emergence, it questions the notions of inevitability that too often surround studies of globalization. Particularly useful for comparison of cases are the models of "world system" developed by Modelski and Thompson (1996, see Devezas and Modelski, 2008). These focus on national policy-driven innovation and processes of European "evolutionary learning" that began in the 1400s. They put into context the models that focus on core–periphery structure as developed by Braudel (1973), or the "world-system" core–periphery model that for Wallerstein (1974) begins in the 1600s. Study of structures of core–periphery in world systems can benefit from added dimensions, improved measurement of network structure, and understanding the effects of periodic crises in terms of historical dynamics.[1]

An unexpected outcome of this survey for issues of policy is that it develops a deeper historical understanding of how certain kinds of exchange systems develop

---

[1] Against determinism. If the static models of purely structural Marxists were correct, many issues of how networks operate in globalization would be deterministic, e.g. intrinsic inequalities in exchange produced by structurally unequal positions. History suggests otherwise. Neither the existence of core–periphery role structure in the world economy (e.g. Reichardt and White, 2007), nor of changes of position in the core–periphery economic structure (Smith and White, 1992) entail that core–periphery exchange structures are everywhere deterministic, say of unequal exchange (as contrasted with the observation that core cities and states usually become peripheral at a later time). Some systems of exchange are circular, for example, as with circular routes and centers in Medieval European trade. Nor do histories of how such arrangements are appropriated by the state tell the whole story of globalization processes and their effects. A deterministic role of the hegemonic state has been highly contested and varieties of pluralistic theory articulated concerning interventions of middle classes, workers, the poor, and others in issues of exchange in the industrial era (Aronowitz and Bratsis, 2002; Tilly, 1996).

Networks in Social Policy Problems, eds. Balázs Vedres and Marco Scotti. Published by Cambridge University Press © Cambridge University Press 2012.

several kinds of inequalities that are inimical to the concept of fair pricing in the operation of market equilibria, even in the absence of economic oligopolies (monopoly, duopoly) and oligopsonies (monopsony, duopsony). These include longstanding militaristic state-policy domination of international exchange, resultant structural inequality in international trade networks, and cyclical events within polities, that in periods of resource scarcity relative to population, create periods of extreme deflation of wages relative to extremes in elite dominance over wealth-generating property ownership.

Consideration of policy alternatives begins with how deeply these inequality-generating processes operate long term, not only historically but in the present day, amplifying the sellout of our future, politically, economically, and ecologically, to shortsighted objectives, a conclusion reached by many authors, most recently Coyle (2011). While Coyle stops short of what to do about making markets moral and towards what ends, I go on to what to do about these problems given their history. If "more money makes people happier because it means they can buy more" (Coyle: jacket) then entails economic growth as the objective, when economic objectives are backed by military policy, population growth and economic growth become incompatible. In these circumstances – of militarized economic policies – population growth is conducive to social conflicts, debilitating population growth, and worldwide impoverishment, and results in less, not more, purchasing power for most people.

The present inquiry problematizes Modelski's notion of a rather invariant type of leading state policy-driven process of "evolutionary learning" that creates and perpetuates core–periphery hierarchy in economic structures. Outcomes in network globalization processes are potentially contingent on different kinds of cycles of economic instability and change that are mutably affected by policies – state or nonstate – embedded in a broader framework of networks of agents involved in trade. In examples of medieval exchange, Arrighi and Silver (2001) analyze trade networks, in terms of a contrast between commercial capital and financial capital, and between multiparty policies on trade (Genoa) versus state policies (Venice). Spufford (2002) offers an analysis of the role of trade, monetization, and commercial entrepreneurship in the transformation of Medieval Europe. I will show that Modelski's data on militarized trade policies in leading economies, from 1450 to the present, entail displacements of commercial ethics in all but community-based trade, replaced by what Jacobs (1992) calls a Guardian ethic of winner-take-all.

History is rife, however, with diverse examples of trade networks in which commercial trade entrepreneurship and ethics dominated exchange, rather than state policy or financial capital. These include commercial capital and innovation developing at various sites include Sung China, Champagne Fairs, the Silk Roads in the Empire of the Golden Horde, nonreligious aspects of the Hanse League, the

English sea captain "Asian malfeasance" trade post-1680, and some of the policies of the European Union.

If there are alternations between commercial and financial capital (pace Arrighi), what causes them? What entails the shift to state-dominated economic policy? Is that shift connected with specifically financial capital? How does this interlock, if at all, with Modelskian phases of innovation that may lead to market dominance of a leading state? Under what conditions do alternative types of scenarios reoccur within the globalization framework that involve bottom-up processes of multi-agent networks within which trade policies emerge? Do these generate periods not only of peaceful trade but also of reduced structural (core–periphery) inequality in the advantages accrued from trade? The range and scope of this chapter opens questions but does not purport to offer law-like solutions.

## 11.1 Scale and temporal cycles in globalization processes

Noting Modelski's political science overemphasis on the ubiquity of state policy as a driver of economic globalization does not entail disagreement with his overall framework, which is discussed later in the chapter. Historical and ethnographic data from periods spanning dozens of centuries and years (Turchin, 2003, 2005, 2006, 2009, 2010; Turchin and Nefedov, 2009; White *et al.*, 2008; White, 2009a) and at different spatial scales, however, shed different light on the periods that Modelski and Thompson (1996) include in their study of "evolutionary learning" in globalization. These, in their view, begin in CE 900 with the invention of printed money in western China. This first period led a developmental process that contributed to the twelfth and thirteenth century early European Renaissance. Some aspects of this early period of globalization are echoed in the present, as this is not the first time that China has been a new center of commercial innovation supplying commodities to the West.

A review of some findings from these data that are relevant to issues of economic globalization may best be examined by starting at the largest spatial scale within continents, then focusing in on smaller regions that are of a size where population numbers create Malthusian demographic pressures of relative scarcity in food resources. White *et al.* (2008) show city-size distributions within regions and inter-city trade networks that interact with the dynamics of inter-regional trade. Here, I show how these interact with the ethical codes and policies that govern trade at different levels.

### 11.1.1 *States and cities (globalization and networks of cities and towns)*

Globalization is a process of expanding the network of trade across cities and also includes exports and imports from interdependent rural zones of cities. Jacobs

(1983) notes that "the power of city markets, the power of city jobs, the power of city technology, the power of city work transplanted out of cities, and the power of city capital ... are the five forces that shape and reshape all regional economies." However, "In the hinterlands of some cities – beginning just beyond the suburbs – rural, industrial, and commercial workplaces are mingled and mixed together. Such city regions are different from all other regions, especially those having the richest, densest, and most intricate economies to be found, except for those of cities themselves."

Jacobs (1969, p. 262) defines a town as "a settlement that does not generate its growth from its own local economy and has never done so. The occasional export a town may have generated for itself has produced no consistent self-generating growth thereafter." What Jacobs forgets is that (1) cities are dependent on their external trade, and (2) some town regions engage in long-distance international trade in addition that indirectly self-generates growth, if we consider the global economy as including long-distance links that are not strictly inter-city. The network view that would enable an understanding of non-city links in long-distance trade is to look at the items that travel along these chains, in terms of the longer or longest distance pairs of endpoints of production and consumption. New manufacturing and trading groups in China, for example, also engage in long-distance links that are not exclusively inter-city: e.g. a peasant ironworker flies to Britain, uses bank loans to buy an inactive steel factory, ships it back in pieces and restarts the factory in China, later branching into autos. Nor was it simply the contemporary Chinese *state* or *cities alone* that fueled China's recent economic development as a new center of commercial innovation supplying commodities to the West. Kynge (2007) argues that it was the recent legalization of private and kin-group entrepreneurship, unleashed but not controlled, in China's recent economic liberalization.

Similarly but long ago, it was not the Chinese tributary state and urban centers that created the early (900–1100 CE) developments in commercial globalization (including mechanisms of credit), but Chinese entrepreneurial farmer-merchant families of the western Silk Road regions who engaged in the network of long-distance Eurasian trade. In work on city-size distributions in world regions, White *et al.* (2008) noted that size distributions typically have a scale-free or power-law log–log tail of upper-size cities, but that convex fall-off in the lower-size regions of the distributions tells us about the relative economic health of the smaller-size distributions. Those small town-size regions with heavy convex fall-off have an advantageous packing ratio that creates many smaller rural "hubs," less space between smaller cities, and (as with the denser populations of cities) more competition for resources. In such regions, individuals are likely to have a higher density of roles and responsibilities than in cities, and contribute to stable if not the most rapid innovation. It is imperative that equal effort be devoted to study of the parameters of smaller-scale towns and settlements before leaping to the conclusion advocated by West (Lehner, 2010, pp. 50, 52) that "modern cities are the real centers of sustainability," "a growing city makes everyone in that city more productive," or that "creating a more sustainable society will require our big cities to get even bigger. We need more megalopolises."

While the world cities – New York, London, and Tokyo – of Sassen (1991) still control disproportionate amounts of global business and international influence through a hierarchy of specialized organizations (HQs and multinationals, finance capital and services, stock markets, giant international law firms), and living costs and individual wealth are greatest in the largest cities (Sassen, 2000), the phenomenal growth of near-instantaneous Internet services in domestic markets (MERS, for example, founded in Reston, VA, which since 1995, processed millions of mortgages, merely through self-nominated agents with no subsequent supervision) or international outsourcing to Bangalore, Hyderabad, Indore, Kolkata, Pune, or Mumbai, has rapidly changed the spatial structure of advanced economic centers, often putting high-end functions together with sites of impoverishment.

Bettencourt *et al.* (1995) argue that the productivity of city economies scales with their size, i.e. that "gross economic production, personal income, numbers of patents filed, number of crimes committed, etc., show super-linear power-scaling with total population, while measures of resource use show sub-linear power-law scaling" (Shalizi, 2011). This entails that resource use (e.g. declining energy use per capita) is more efficient with larger city sizes, but productivity is increasingly higher. Shalizi (2011, p. 7), however, shows for these super-linear relationships that neither "gross metropolitan product nor personal income scales with population size for U.S. metropolitan areas", but rather result from failure to use per-capita measures, which was also the case for the other "superlinear" variables. Intriguingly, Shalizi (2011) also finds that "the appearance of scaling with city size is in fact explained by the correlation between size and occupational structure."

"In his more pessimistic moods", and extrapolating from scaling results that may (or may not) show increased productivity in cities as benefits of growth, writes Lehner (2010, p. 50), West "knows that nothing can trend upward forever ... And so growth slows down. If nothing else changes, the system will eventually start to collapse." Yet (p. 54) "West describes cities as the only solution to the problem of cities," i.e. through innovation. "Once we started to urbanize ... We traded away stability for growth. And growth for change." (Note the lack of differentiation between economic growth, growth of city size and functions, and regional population growth). The advantage of cities, "as West put it" is that they cannot be centrally managed, "and that's what keeps them so vibrant." "It's the freedom of the city that keeps it alive." That freedom may be enhanced by the further variable of multi-cohesion of collaborative economic network ties that heighten productivity (White, 2009a). Katz (2010) argues that if larger cities are more productive per capita (or not: dubiously as well, notes Shalizi, they may walk faster and burn more energy per unit time), this is less possible outside urban and developed peri-urban economic zones, given the higher quality of primary interpersonal networks rather than ties by Internet connections. Defects of Internet impersonality are illustrated by the computer generated increase in property tax mailings that generated California's 1968 tax revolt. Advantages of long distance low energy links are illustrated by the Egyptian revolution of 2001. Bettencourt *et al.* (2007) have now gone on to study statistical distributions of innovation and productivity of corporations, having "an average life span of 40–50 years," and "decline in productivity per employee" that makes for "increasing vulnerability to market volatility."

My argument, however, is that once trade policy comes to be controlled by national-interest mercenary or militaristic policies on trade, in which aggression has undercut fair pricing, rather than those that favor benefits of fair trade – between whatever parties at whatever distance – and fails to represent the interests of the trading groups themselves, on an equal footing, then the market economy becomes tilted by relations of unequal power (as in Aristotle's theory: price as altered by inequality of status). Mercenary ethics of trading impose unfair pricing to the detriment of counterparties. Alternatives include, among others, free trade, policies of self-regulation by traders, complemented by the anthropological concept of generalized exchange under an ethic of giving back to others (or that which benefits another benefits all, but not necessarily in the short term), and pricing by full replacement cost.

The city builds on productivity, needed for its dependence on trade, and pays for creativity, which it cannot survive without. In addition to collaborative multi-cohesion that may provide "fair trade," in the use of "mercenary" trade backed by aggressive state-policy pricing imposed on others, cities are paying for their existence well below "fair prices," which imposes low profits to those outside cities and regions with which they trade. The city model of Bettencourt *et al.* (2007) fails, at best, without taking into account more global economic relationships that may entail unequal exchange.

Modelski and Thompson (1996), however, chart the wealth and innovation of leading cities at the expense of competing regions with national trade policies. These policies were aimed at killing off or absorbing through conquest the leading cities of other regions, i.e. following an ethical morality observed by Jacobs (1992) that shuns trading for exertion of prowess, obedience, discipline, tradition, hierarchy, loyalty, fortitude, fatalism, deception for the sake of the task, ostentation, largess, exclusiveness, emphasis on honor and leisure, and the taking of vengeance.

## 11.1.2 *Ethical codes and systems: Jane Jacobs*

Jacobs (1992) argues that two fundamental and distinct ethical systems, both valid and necessary, among others, govern the conduct of human life. One is a commercial code that shuns force in favor of competition, stresses efficiency, thrift, industriousness, inventiveness, novelty, initiative and enterprise, voluntary agreements, respect for contracts, honesty, optimism, collaboration with strangers and aliens, and dissent for the sake of the task – i.e. valuing transparency. The other is the political Guardian code – descended perhaps from feudal landed estates – that shuns trading for exertion of prowess, obedience, discipline, tradition, hierarchy, loyalty, fortitude, fatalism, deception for the sake of the task, ostentation, largess, exclusiveness, emphasis on honor and leisure, and taking vengeance.

The shift from Asian-based leading sectors of the commercial-code type, as described by Modelski and Thompson (1996, pp. 171, 191), to European leading sectors, takes place with the rise of the Mongols of the Golden Horde starting in 1190 that, in creating greatly increased East–West trade from Asia to the Black Sea, favored Genoa over Venice for nearly a century due to Genoese control of the Black Sea trade. Genoa had evolved from pirate protection rackets hundreds of years earlier into a decentralized commercial market composed of several thousands of extended families. The Genoese from 1190–1250 shipped these Asian goods north to exchange goods at the Champagne Fairs and from 1250–80 enlarged their Black Sea trade further (Modelski and Thompson, 1996, p. 191). Thus, "In the thirteenth century, the locus of innovation shifted to the Mediterranean, led first by Genoa, and soon followed by Venice."

Genoa was the last decentralized leading commercial market city in Modelski and Thompson's "leading sectors" of globalization, one in which competing families shipped their members out to distant ports to set up independent trading posts, and moved goods and capital between them. Genoa became, by 1290, the richest port in Europe (Spufford, 2002, p. 26), and trade route network analysis shows that it was highest in betweenness centrality at this time. Betweenness centrality favors success in mercantile trade.

Withdrawal of the Mongols of the Golden Horde from their protective role towards the Silk Roads in their conquest of the Southern Sung (1279), however, undercut Genoa's routes to Asia. The Venetian innovation of enlarging their galleys from 20 to 150 ton cargo capacity between 1300 and 1320 (Modelski and Thompson, 1996, p. 190) also allowed them to undercut the Genoese access to Black Sea ports (1320–55), and expand their pepper trade (Modelski and Thompson, 1996 pp. 187, 191) through Mamluk Red Sea middlemen. Although Genoa controlled the Black Sea trade of slaves to the Egyptian Mamluk Sultanate, they were unable to break Mamluk control over the Red Sea route to Asia, which was used by the Venetians in the 1290s. "In this same decade, a large contingent of

Genoese were building galleys in Persia so that the Mongol ruler of Persia might interdict the Indian–Egyptian trade by blocking the Red Sea entrance" (Modelski and Thompson, 1996, p. 187). Genoese market ethics were competitive, but within a commercial equilibrium market framework, rather than coordinated by state policy.

Venice was a state-policy economy with charter shareholder corporations that contracted with trade partners, sent family members of shareholders to sea on a corporate vessel, but family members returned to the corporation to divide the goods and profits from the voyage. Venetian state-policy strategies won the competition with Genoa and dominated the Asian trade from 1300 to 1430.

What Modelski and Thompson (1996) describe in their treatise on "evolutionary learning" after 1430, are the results of European national policies involving Guardian–conquest codes imposed on the international economy. This code has survived to become the dominant force today, in competition with marginalized commercial codes of ethics, which so dismally collapsed in the current meltdown of financial innovation, with its policy-driven nontransparency and insider political and banking machinations (Schweitzer *et al.*, 2009a,b). Contemporary Euro-American financial institutions, for example, act on Jacobs' Guardian code of punishing whistle-blowing and give vast rewards for loyalties that do not betray endemic illegal practices.

## 11.2 K-waves of innovation with state-driven political codes

Devezas and Modelski (2008, pp. 15–16) define evolutionary learning as a part of evolutionary theory, which must specify mechanisms of change (Giddens, 1994). In the context of globalization, Modelski and Thompson (1996) view *evolutionary learning* in terms of successive regions that become dominant through competition, usually punctuated by warfare as the strength of a leading challenger comes to match that of the current leading contender. The breakthrough period is one that gives the political clout to organize challenges to current leading states. Phases of agenda setting (pp. 7–8, 53, 67, 222), coalition building (p. 106), macrodecision (p. 68), and execution, correspond with early forms of Kondratieff waves (K-) waves of innovation, basebuilding, networking, breakthrough and payoff. They conceive of each as "a response to a specific cluster of innovations", and see these K-waves as following a recurrent pattern of four-wave phases within a region that becomes economically dominant, from basebuilding to networking to breakthrough to payoff.

Their view divides the formative period of globalization – the Sung breakthrough through the Early European Renaissance and its expansion of the nautical/commercial revolution in Europe and the Mediterranean – and the state

Euro-capitalist formative and consolidation periods that begin with the Portuguese circumnavigation of the Afro-Eurasian continent. The latter process of globalization began in earnest by 1430, through Portuguese state policies, and opened explorations that led to the New World Portuguese and Spanish conquests of the next century.[2]

Surprisingly, Devezas and Modelski (2008, pp. 31–32) describe for post-1430 market evolution a "predominantly commercial nature of the leading industries endured ... from the fifteenth through to the eighteenth centuries," after which "the emphasis on the character of innovations shift[ed] to industrial production." In actuality, however, the period of European domination of trade from the fifteenth through to the eighteenth centuries did not represent a commerce-based economy, but militarily backed state policies competing for domination of trade, with the following period only euphemistically called industrial and then post-industrial capitalism.

### 11.3 Rise of a "Guardian" view of economics and globalization: invasive European global hegemony in Asian trade

Modelski and Thompson emphasize, in their broad global comparative study, that the histories of state-driven trade policies oriented towards trade-dominance and warfare, which emerged in Europe in the 1400s and continues to the present day. Even today the USA. has trade wars with Brazil over its agriculture subsidies that violate agreed-upon WTO regulations. Under what conditions do the dominant players in trade networks and policy formulation tend to operate top-down, as in state policies aimed at dominance, or bottom-up, as with multiethnic trader networks and multilateral or international organizations aiming to optimize the benefits of cooperation?

The type of state-policy globalization that began in 1430 in Europe, in the Modelskian view (Devezas and Modelski, 2008, p. 32), was "already an energetic and fervent activity, with leading sectors clearly commercial in nature." But for Portugal, the Netherlands, Britain, and the USA. as leading sectors, international trade was rarely governed by commercial merchant codes. Instead it involved state-driven policy objectives for domination of the world economy, beginning with basebuilding for commercial and political networking, aimed at breakthrough to a nationally based economy that, by means of global dominance, then leads to economic payoffs and the political power to wage successful wars against competitors.

---

[2] By 1510 the richest city in Europe was Lisbon, where bullion was pouring in from the New World. By 1600 it was Antwerp, the banking city with the greatest global flow betweenness in terms of European trade. Flow betweenness, in exchange networks, favors the dominance of financial centers.

This created a cyclical dynamics with few commercial-ethic interludes outside the policies of leading sectors and states.

The first instance of this type of trade, planned and executed by the Portuguese kingdom, drew from the example of Genoese trade incursions in northwest Africa (1290s) which were aimed at securing gold to finance circum-African trade routes to Asia, as an alternative to the Mamluk-blocked routes through the Red Sea and the Persian Gulf. Following the Portuguese defeat of the Moors (Algarve 1249, and the 1415 counterattack on Ceuta, Morocco, which marked a start for European colonialism), Portugal's King John I organized scientific and policy conferences to assist the state in developing strategies and infrastructures to round the Horn and to invade and conquer Asian markets.

The project of the Portuguese as the European system-builders of the fifteenth century (Devezas and Modelski, 2008) began in 1420 with the technological advance, by 1440, of flat-bottomed caravels capable of navigating in shallow waters and steering into the wind, engaging the challenge of getting ships around the Guinea Coast (*da Guiné*, Guinea Return) in order to gain access to gold from the deposits of West Africa (*Volta da Mina*, circle of Mine Return). These was further development by the 1440s of the quadrant and progress in scientific conferences and engineering (1480s conferences of King John II) and guidebooks for navigating the coast by the use of the *balestilha* (cross-staff) to measure latitude from the Sun's angle of altitude. By the 1480s caravel artillery was installed on fleets of larger, sturdier caravels capable of long ocean voyages that pushed around the Cape of Good Hope. Within the next 50 years they had established the Indian Ocean Spice Trade, developed huge galleons for war and transport, and enlarged a network of military bases from the west coast of Africa to the rich coast of India. The Portuguese imposed a protection racket on Asian traders as the price of not destroying their ships. Prior to this first maritime invasion of Asian trade networks by the Portuguese, Asian sea trade had been peaceful.[3] The third of the maps in Figure 11.1 shows a dashed arrow that represents the extent of progress in the planned Portuguese route to India in 1450.

The original talk for this chapter (White, 2009b), at the founding meeting of the Center for Network Science in Budapest, used a slideshow of network graphs showing maps of Eurasia from 900 CE to the 1950s, overlaid with overland trade networks between the largest cities listed by Chandler (1987), at 50 year intervals, and with shipping routes from Europe to Asia annotated as to effects on European

---

[3] "The peaceful Asian sea trade network, which peaked during the Chinese Ming dynasty, was disrupted by the coming of European traders in the 16th and 17th centuries, which brought the establishment of the global economy" (Erika Rempel HST 101 February 29, 2008 – The Clash of the Asian and European Sea Trade Networks) is referring to early Portuguese conquest, three armed fleets headed by Vasco da Gama 1497–1501, sailing to India.

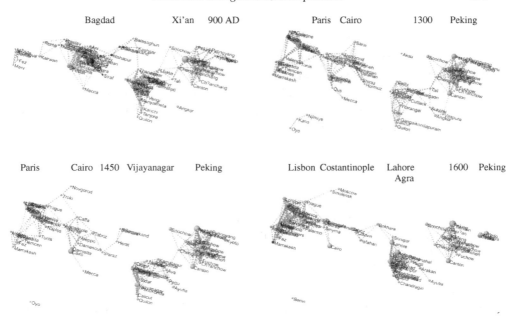

Figure 11.1. Sample trade graphs for Eurasia. Dotted lines evoke major invasions.

trade of military dominance. The original images of these maps can be seen at http://intersci.ss.uci.edu/wiki/ppt/economic_networksCEUøg.ppt.

The maps aimed to show how land-based multi-cohesion in trade routes (favoring price-equilibrating markets) breaks up into noncohesive segments when warfare occurs over trade routes, maritime and/or inland, or sea-route domination by leading powers. Four periods appear on the maps. The most consistently cohesive are (1) those up to 1300 CE, dividing into two geographic segments, W. Arabs versus Others (950–1000 CE), and Mediterranean versus East Asia (1100–1300), and (3) from 1650–1750, only China is separated from the rest. The least trade-cohesive periods are (2) those from 1350–1600, which divide into three or four segments (Euro-Med versus Mid-East versus India and China 1350–1400), W. Eurasia versus India versus China (1450), with India-China consolidated in 1500, but separated again in 1550–1600, and (4) from 1800–1950, starting with the British–Chinese opium trade and trade wars and a high conflict period generally, trade group numbers are unstable. Periods 1 and 3 have a similar profile of city scaling coefficients (White *et al.*, 2008), rising from devastated to over-integrated indices of city size (the latter means too many largest cities for the same region) for Eurasia generally and for China only, from normal to over-integrated in period 2, and fluctuating wildly from 1800 forward. The earliest periods of globalization (900–1250) are those with greater trade cohesion (N = 1 larger markets, east and west), favoring "fair trade" prior to European domination. Period 2 and 3 boundaries (1300, 1600) initiate disruptions in rate of population growth (McEvedy and Jones, 1978).

There is a reasonable dynamic correlation of inter-city trade connectivity in the Eurasian maps and city-size distributions by region. Lesser connectivity ($N = 3$–$4$ separate cohesive "markets") inveighs against fair global pricing and results in erratic city-size fluctuation, ending in periods of urban collapse. Restoration of larger trade connectivities ($N = 2$) restore Zipfian city hierarchies in periods that may end in over-integration.

Devezas and Modelski (2008) give a brilliant exposition of their concept of "evolutionary learning" with respect to globalization, as Portuguese state policies were implemented, step by step, usually in five-year increments. This is not a spontaneous or bottom-up evolution, however, but one organized by state policies. My view is that the state policies of globalization through conquest, which proceeded from the Portuguese/Spanish conquest, to the English and Dutch, to the English and American, need to be rethought. These are not fair market-price equilibrium economies. They are prima facie unequal exchange. They do not provide the benefits of free trade and processes leading to price equilibria promoted by economic theory. The term "market" in these cases serves as window-dressing for various forms of tribute-based international exploitation that have invaded many domestic economies from overseas.

The Portuguese invasion was the first state-driven globalization policy based on conquest. It marks a startup of "the history of the world system, within just one embracing concept" (Devezas and Modelski, 2008, p. 12), for which Modelski and Thompson employed the term "leading sectors." Modelski and Thompson's "world system" startup in the fifteenth century is thus at the root of Braudel's (1973) and Wallerstein's (1974) "world-system" emerging in sixteenth century Western Europe and the European expansion into the Americas, in which "economic exchange between core and periphery takes place on unequal terms: the periphery is forced to sell its products at low prices but has to buy the core's products at comparatively high prices" (Wikipedia: Wallerstein Theory). What Modelski and Thompson (1996) and Devezas and Modelski (2008) have revealed, however, is that Wallerstein's and Braudel's emphasis on economic core–periphery structure turns out to be generated by explicit state policy aimed at trade conquest or market tribute, which undervalue fair market prices for the conquerors.

Alternative policy formulations can be sought that replace the explicit violence and warfare involved in European polities that established Euro-globalization. Postwar decolonization was not sufficient to level the playing-field of trade (Smith and White, 1992). To see how the economic playing-field might be leveled we need to dig further.

## 11.4  Commercial-transparency code examples

The policy questions examined here, in relation to Turchin's studies of Malthusian cycles with breakouts of extreme inequality and the K-wave cycles of Modelski and Thompson, concern the role of regulatory systems and their cultural contexts (Turchin, 2003, 2005, 2006, 2009). Examples of more transparent commercial economies given here begin with a focus on medieval globalization – the invention and spread of paper money, credit notes and expansion of trade, free of tributary control. The juxtaposition between policy – usually considered to be state policy – seems oxymoronic in the context of medieval globalization, in that many periods were not state-driven in trade or monetary policy. This hints at policy possibilities under international law rather than states directly, which may only need to ratify international law. I ask: what would it mean to have states minimize their own conflicts in favor of a strong but level playing field for trade, such as happened frequently in the Medieval Chinese or European periods of renaissance? What would it mean if state policy towards globalization and trade strategies were not conducted with a goal of defeating or eliminating competitors in globalizing markets? Market capitalism emerged before state and financial (putative) capitalism, and I argue that the intrusion of Western state militarist trading into relatively peaceful Eastern markets in the 1450s began a chain of historical events that was a negative form of "evolutionary learning" in the processes of network globalization, neither a model for the present nor for future policy.

### *11.4.1  Sung (900–1190) Silk Road trade and the commercial code ethic*

Trade on the Silk Roads fluctuated for a millennium (e.g. 300 BCE – 900 CE) as alignments changed among the tributary states of Eurasia. Rome and the Eastern Empire of Constantinople organized vast amounts of trade but pre-Medieval trade was predominantly tributary and administrative and not free merchant trade. Modelski and Thompson (1996, p. 132) credit Sung China (Northern Dynasty 900–1127) with the breakthrough globalizing invention of a national currency-based economy, in four stages: printing and paper, including paper money, the basebuilding stage (900 CE – 990); national market formation, the networking phase leading to globalization (990 CE – 1060: onset of globalization *c*. 1000); the construction of a fiscal/administrative framework, the breakthrough stage facilitating Sung globalization (1060 CE – 1120); maritime trade expansion, and the payoff stage of the globalization process (1120 CE – 1190). Paper money developed from "a surge in the use of printing and paper industries" (Modelski and Thompson, 1996, p. 131), the use of paper notes of credit by eastern Chinese merchant families at the terminus of the Silk Roads, out of which banking families evolved in the region, and the adoption of paper money and notes as the basis of national economic transactions.

### 11.4.2 Champagne fairs (1190–1280) → Paris: fair trade regulations

The Champagne Fairs of the High Middle Ages (twelfth–thirteenth century CE) offer an example of "merchant law" domination of the commercial and banking relations that came to regulate trade between the Mediterranean and the north European lowlands. Operating at a north–south frontier, these fairs equalized the power differentials between cities and states and allowed inclusion of more distant parties on equal terms. Later the exchanges came to be executed by brokers rather than primary representatives of the merchants exporting their goods (Spufford, 2002, p. 148), which favored migration to Paris, capital of the most powerful European state. The French king from 1285 to 1314, Philip IV, Count of Champagne and King of Navarre, undercut the fairs with his ascension to King by moving the Champagne Court to Paris (also moving the French mint to Troyes), and by making war on Flanders, from 1297. Peace in his domains enabled the first commercial sea routes from Bruges to newly conquered Cadiz, open also to Genoese traders. Effectively, "merchant law" trading was dead from Flanders south by 1300, displaced to the north by the church-based merchant relations of the Baltic Hanse League.

### 11.4.3 Silk roads and Mongols (1226–1369) – understanding multiconnectivity in exchange systems

In their typology of the K-waves of the political powers that become dominant in the Silk Roads system, Modelski and Thompson (1996, p. 148) leave out consideration of the multiconnected international structure of the core network of exchange routes: the four-cycle system connecting Northern Sung to Constantinople to Cairo to Tanjore back to the Sung, land routes to the north connecting to sea routes to the south. The network theory for such cycles is that a cyclic structure facilitates the emergence of competitive pricing and cooperation. "Plott notes that, in a tightly controlled market (monopoly/monopsony; duopoly/duopsony), economics has no theory of prices (personal communication, 2007). This is because, in a market of exchanges, price equilibrium occurs only with bids from multiple buyers and multiple sellers."

Cohesive multiconnectivity without a chokepoint or dominant controlling center, but with alternative routes by disjoint paths in a network of trade describes a network that facilitates the emergence of equilibrium pricing. From pages 149 to 204 of Modelski and Thompson, in the chapters on Sung China, Renaissance Italy, and on the startup of later periods, the Mongols were major players. Italian K-waves in the Mediterranean (Modelski and Thompson, 1996, p. 206),

1204–1453 CE, are shown alongside a column for transition crises in the "Mongol World," which at its maximum, had territory nearly equaling that of the British Empire. The column includes: the wars of Ghengis Khan (1206–1226) and pacification of intracontinental trade routes; the Mongol sacking of Baghdad (1258); the conquest of Southern Sung (1266–1279); the expulsion of the Mongols from China (1367); and the conquest of Persia by Timur, whose subsequent defeat of the Golden Horde led to total collapse of the Mongol Empire after 1369.

Prior to their expansion, the Mongols had a pre-monetary exchange economy of herd animals in bridewealth (horses, camels, sheep, and other livestock) in exchange for brides who provided fertility for stocking extended family, daughters, and warriors. If wealth is defined as social or economic capital that is productive of further wealth, then pastoral wealth against bridewealth exchange in fertility, as in tributary and monetary economies, generates population growth in some periods, due to wealth-production, later culminating in resource scarcity relative to population which results in growing inequality of laborers and the wealth-owning elite and their warriors.

The policy-relevant aspect of the brief Mongol administration of the Silk Routes was that they cleared the trade routes of obstacles and promoted trade. Economies burgeoned at the intersection of ethnicities along the routes. At one point, Samarkand was the most cosmopolitan and luxurious capital of any region in its time. Although this brought vast riches to Mongol nobility, the hardship of administration when subject people were often devastated by the Mongol war machine brought the third and fourth generation clan rulers to convert their policy from extortion and punishment of subject groups, often destroying their fields and resources, in consideration of "the enormous tax benefits of permitting peasants to cultivate their lands" (Modelski and Thompson, 1996, p. 184), and consequent focus on conquest of the Sung (1266–1279), which created later conditions for transmission of the Black Death (1346–1353) from dense Asian populations to the Black Sea and the West.

Collapse of the Black Sea trade with Timur's conquest of Persia and disintegration of the Mongol Empire (1369) led again to reorganization of trade routes, with the northern and Persian Gulf routes replaced in importance by the Red Sea route. Increasingly sea trade replaced or challenged overland trade.

### 11.4.4 Sea captains (1620–1824) – British non state-driven trade: democratic capitalism

A third type of data examined in my original Budapest talk (White, 2009b) was time series network data on 4,572 voyages by English traders of the East India Company (Erikson and Bearman, 2006, abstract; see inset below) from 1620 to

1824, showing "dense, fully integrated, global trade networks" that reveal free-trade market globalization prior to the industrial globalization of the 1800s. Erikson and Bearman's purpose is to show that the "origin of modern economic markets" (i.e. high volume free trade with market price equilibria) emerges in an unexpected way, from the malfeasance of ship captains paid by the British East India Company (EIC) who, instead of returning from Asia with commissioned goods, would intentionally "miss" targets for return voyages (given the calendar of the trade winds) and stay in the Eastern trade orbits for an extra year, trading on their own account and making fortunes of their own, while satisfying the formalities of their contracts with the EIC.

My purpose is this example is to show a form of capitalism that can emerge with free trade and price equilibria that is not biased by state policies. State policy governed EIC trade in 1620–1624 (as shown in inset A of Figure 11.2), but in competition with the VOC (Dutch East Indies Corporation) the "persistent lack of resources and competition with the VOC led the EIC to adopt cost-cutting measures – abandoning the intra-Asian trade, abandoning surveillance of employees, leasing ships, and insufficiently supplying factors [sea captains] with bullion. Part and parcel of this retrenchment was the adoption of a strategy of employee appeasement as a solution to the principal–agent problem" (Chaudhuri, 1978). The consequent reduction of EIC trade is shown in inset B for 1660–1664.

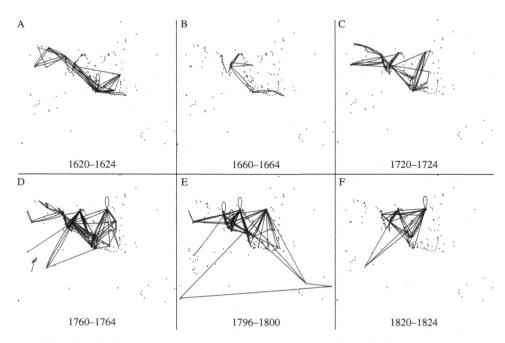

| A | B | C |
| 1620–1624 | 1660–1664 | 1720–1724 |
| D | E | F |
| 1760–1764 | 1796–1800 | 1820–1824 |

Figure 11.2. Development of private trading in the English Eastern trade in snapshots of six four-year periods (Erikson and Bearman, 2006, p. 212).

Then, "[b]etween 1667 and 1679, in an effort to routinize the home market and reduce exposure in the East, the EIC gradually withdrew from the country trade. By 1680 the EIC formally abandoned intraregional trading, which was left to individuals" (Marshall, 1993). By the time of the "Glorious" peaceful democratic revolution of 1688–1689 (Pincus, 2005) Protestant and Catholic interests were reconciled, while sea captains and merchants brought new capital assets to play that displaced the hegemony of landed estate-owners. The wealth of the captains and merchants was acquired in the decade of sea captain malfeasance (private trading), which also benefited merchants at the English ports.

This example shows how transformative unintentionally instigated private trading was in, bringing about a change in fortunes of the aristocratic landowning classes vis-à-vis the merchant and entrepreneurial class, whose wealth was as a result of the potential for profits in free-trade and high-volume market equilibrium. By 1689, landownership was displaced by an early form of parliamentary democracy as the basis of English state power.

> After 1680, when the EIC had withdrawn from the country trade, commercial opportunities within the East were numerous. Although the endogenous country trade in the East was significant and commercially sophisticated throughout the 17th and 18th centuries, it was fragmented into largely disjointed markets. This had not always been the case; Eastern trade systems were previously much more tightly integrated, and *the penetration of Europeans contributed to a process of fragmentation already underway* as a consequence of the decline of the Muslim Sultanates and the withdrawal of Chinese-sponsored foreign trade. By 1680, the Eastern trade system was dispersed and loosely jointed, split between markets serving China and those centered on the Indian subcontinent, further divided among isolated merchants loosely tied to geographic bases. For example, Gujaratis dominated the trade across India and Bugis operated in the Indonesia archipelago. Between 1681 and 1764, structural holes resulting from this regional clustering were bridged by English captains free riding on EIC resources in pursuit of private profit (Erikson and Bearman, 2006, p. 201; my italics)

The period of English free trade in Euro-Asian markets eventually displaced the type of Portuguese militaristic conquest trade in the East (carried over to the New World by Iberian conquests) but ended with the EIC coming to rule large areas of India (often through profiteering agents who violated company policies; as today, it was difficult for states to regulate in favor of free trade), gradually abandoning commercial pursuits in favor of company rule of India, which was consolidated in 1757, after which the intensity of EIC trade began to dwindle (insets D–E–F of Figure 11.2).

The effects in India of colonial VOC and EIC trade enforced by state-policy and military rule, up to independence in 1949, need not be rehearsed here. My presentation from the Budapest conference went on to evaluate the harm to the

Chinese economy following the invasion of the British and the opium trade. Continuance of militaristic trade policies of leading sector countries, however, operate to the present with financial codes of ethics to "maximize profits by any means necessary."

## 11.5 Reasons for emergence of financial codes of "Guardian" ethics

Why do the state-policy militarized economies of dominant European powers from 1430 to the present fail to utilize the "commercial code" of markets, with equilibrium-fair prices? Is it necessary that national-policy "Guardian codes" of international trade typically morph into a bogus form of financial capitalism?

One answer is easy to see: the technology of China, accumulated from over a thousand years prior to the tenth century, plus Asian exports of the Sung and Mongol periods (the "spice" trade, inclusive of a host of material technology and luxury goods), offered in trade over the Silk Roads, represented a type of exchange in which the demand for precious metals at the Asian end (gold and silver) coincided with the very metals that had been used to back currencies in the West for thousands of years. This was transformative for Europe in three ways: **firstly**, the feudal nobility of Europe could extract bullion through mining and African trade to pay for luxuries and innovations from Asia. **Secondly**, the innovation of credit notes and paper money from China enabled royalty and nobility to use paper money as the currency to pay retainers to manage estates and palaces, while the lords built houses in major cities such as Paris that became centers for merchants with new luxury items for sale from the East, ultimately in exchange for bullion. Wages paid in debased metal or paper currency allowed workers to move to cities, proximal to merchants, and to take up crafts and local industries. **Thirdly**, as capital cities emerged under the pressures of different currencies, state policies in capital cities were oriented to restoring monetary balance, especially those of bullion, and supporting banks that had maximal global flow centrality (as defined by Freeman *et al.*, 1991). The finance capitalism that first emerged in Venice, which looted Constantinople for its wealth in 1204 and outcompeted the Genoese by 1300, emerged in a location that was flow-central (see Freeman *et al.*, 1991) in this period of Asian trade. This is the picture that Spufford (2002) gives of Medieval Europe. But as Europe transformed from a merchant-capital economy centered on the Champagne Fairs to a finance-capital economy centered after 1300 on Paris, with sea routes from Genoa to Holland and England and the religious Guilds of the Hanseatic League, the Dutch cities became the flow-central financial centers of internal European trade. The picture that Spufford (2006) gives of financial centers in Medieval Europe is that "they could only grow up on a foundation of industry and commerce, they could survive for some time after industry and trade had moved elsewhere."

The Venetian reliance on middlemen for Eastern Trade was outcompeted by the Portuguese construction of the first direct routes to India, which also required tremendous finance capital. Trade corporations (EIC, VOR) supported by the states of leading sectors and national trade policy were also supported by (and caused problems for) finance capital of banks and states.

The problems of European state-managed finance capital also created problems that made warfare possible due to massive loans with interest payments or political promises, which motivated warfare to restore state finances, as in France. State finance capital and state-policy management of trade were thus closely linked. They also link to the more than century-long cycles and their phases, linked to long economic (Turchin) cycles that include population growth (which may be enhanced by international trade with unfair exchange benefiting larger cities) that lead to Malthusian shortages of food and hyper-inequality.

### 11.6 Hyper-inequality as an effect on states and trade policy

There are as yet few cases where structural-demographic analyses (Turchin cycles) have been done of nations in the period in which they are leading-sector polities, the exception being England in the period 1450–1800, shown in Turchin's (2005, p. 59) graphic of British structural-demographic oscillations in Figure 11.3, and his phase diagram in Figure 11.4. This period includes Modelski's (2001) K-waves (1640–1740) in Britain's phase I as a leading economic sector. The overall cycle of the Turchin graph, however, includes the leading economic-political sector periods for Portugal (1430–1520), Dutch Republic (1540–1610),

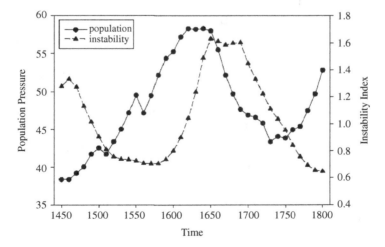

Figure 11.3. Malthusian scarcity/hyper-inequality and civil instability (Turchin, 2005).

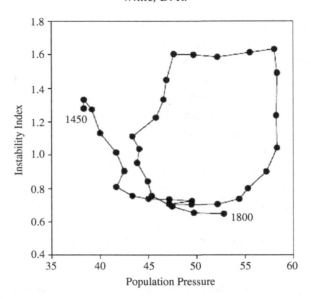

Figure 11.4. Phase diagram of counter-clockwise time dependency in the inter-action between Malthusian scarcity/hyper-inequality and civil instability; "next directions" in the dynamic are predictable (Turchin, 2005).

Britain I (1640–1720), Britain II (1740–1820), in which four K-waves (Agenda, Coalition, Macro, Execution) occur on an average of 80-year cycles (Modelski and Thompson, 1996, p. 8). Modelski and Thompson (1996, p. 111) consider the "high-growth" second K-wave of economic coalition building period predicted from their K-waves theory as 1660–80 (developing the overseas tobacco, sugar, and Indian textile imports), and their observed growth peak as 1670, which accords with my discussion of Erikson and Bearman's (2006) findings (Figure 11.2) about expansion: the sea captains and merchants who brought new capital assets to play that displaced the hegemony of landed estate owners. This decade corre-sponds to recovery from the height of the instability curve in Turchin's graph, in Figure 11.3, as scarcity in the population/resource ratios producing hyper-inequality begins to fall. It anticipates the English "glorious revolution" of the 1690s that coincides with Modelski and Thompson's (1996) "macro-decision" period and global war in which democratizing Britain defeated the autocratic French King Louis XIV (1688–1702). Modelski (2001, p. 79) views the long cycles of global politics as having periods of approximate 120 years and cycles of democratization as periods of *circa* 240 years, while the Modelski and Thomp-son (1996) bursts of four K-waves have an average period of 50–60 years (but in these examples, 80 years). While there might be regular period-doubling in these different kinds of cycles, also major discrepancies. Regular fit in timing of Turchin cycles (which are irregular) and K-waves cannot be definitively established.[4]

What these results suggest from the viewpoint that I have taken, although merely suggestive, is that hyper-inequality (which peaks in the British Turchin cycle, *circa* 1640, when landed aristocracy dominate) is an internal instability that leads to state-policy changes. In this case the EIC commercial trade to India was *withdrawn* by the state in the period 1667–79, replaced by individual initiatives in trade by sea captains, which undermined the dominance of landed estate owners. The Catholic King James II, allied with the absolutist monarch Louis IV of French, attempted to reinforce in battle the succession of a newborn Catholic son against the previous rights of his older Catholic daughter. The recruitment by the British Parliament of royal family member William of Orange and his Protestant wife as Monarchs, and the defeat of James in the battles of 1688 were a continuation of the civil conflict resulting from the period of hyper-inequality. Whereas Turchin and Nefedov (2009, p. 107) argue that the "geopolitical environment apparently played a minor role during this cycle. Although it was the invasion by the Dutch Stat holder, William of Orange, that precipitated the Glorious Revolution, the success of it was entirely due to internal factors," i.e. to conflicts playing out in the structural-demographic cycle. The K-wave and structural-demographic cycles clearly differ in periodicity and subphases, but this example illustrates how they may interact, in this case with a phase of hyper-inequality affecting the conclusion of a major social revolution that had long been simmering and producing a major shift, even if only for a time, to a more decentralized commercial trade ethic.

## 11.7 Policy implications: Arrighi and Silver's view of hegemons

When we speak of leadership in an international context, the term [leadership] is used to designate two different phenomena. On the one hand, the term is used to designate ... that by virtue of its achievements, a dominant state becomes the "model" for other states to emulate and thereby draws them into its own path of development (see in particular Modelski, 1987; Modelski and Thompson, 1996). But to the extent that emulation is at all successful, it tends to counterbalance and hence deflate rather than inflate the power of the hegemon by bringing into existence competitors and reducing the "specialness" of the hegemon ... "leadership against the leader's will," ... borrowing an expression from Schumpeter ..., is always present in hegemonic situations (Arrighi and Silver, 1999b, p. 27)

---

[4] Modelski and Turchin agree that the basic time-scales at which K-waves and structural-demographic effects interact internally are generational, 25–30 years. For Turchin, structural-demographic cycles are not regular as are K-waves, and they cease to apply when interrupted by external events such as war. Being regular in one case and irregular in the other makes their joint timing irregular. K-waves are defined by states while Turchin's cycles are defined only for regions large enough and sufficiently well bounded to have little in- and out- migration so that demographic pressure relative to resources acts (with "sufficient statistics") in a regional not a polity context. Modelski (2000) provides a full explication of wave phenomena that include the economic, political, and behavioral patterns proposed by Devezas (2001).

On the other hand, the term leadership is used to designate the fact that a dominant state leads the *system* of states in a desired direction and, in so doing, is widely perceived as pursuing a general interest ... [T]his is what we shall take as the defining characteristic of world hegemonies (Arrighi and Silver, 1999b, p. 27)

Contrary to the Modelskian view, for Arrighi and Silver (1999a, p. 271), the

... process of globalization of the European-centered world system has *not* proceeded along a single developmental path within which hegemonic states rose and fell. On the contrary, the systemwide expansions under the leadership of each hegemonic state culminated in a crisis and breakdown of the system. Expansion resumed only when a new hegemonic state opened up a different developmental path, reorganizing the system so as to solve the problems and contradictions encountered along the path opened up by its predecessor.

Among their five concluding propositions are several views that support some of the arguments here, arrived at independently:

[1]: 272 The global financial expansion of the last twenty years ... is the clearest sign that we are in the midst of a hegemonic crisis ... that will end more or less catastrophically, depending on how the crisis is handled by the declining hegemon.

[2]: 275 The most important geopolitical novelty of the present hegemonic crisis is a bifurcation of [concentrated] military and [dispersed] financial capabilities that has no precedent in earlier hegemonic transitions. The bifurcation decreases the likelihood of an outbreak of [global] war ...

[3]: 278 Unlike the global financial expansion, the proliferation in the number and variety of transnational business organizations and communities is a novel and probably irreversible feature of the present hegemonic crisis ... and can be expected to continue to shape ... disempowerment of states.

[5]: 286 What lies in front of us are the difficulties involved in transforming the modern world into a commonwealth of civilizations that reflects the changing balance of power between Western- and non-Western civilizations ...

The brief reviews here of commercial markets and ethics tend to agree with Arrighi's view on the possibilities of periods in which exchange is not state-policy regulated, which is assumed by Modelski and Thompson, but involves broader multiparty interactions. Arrighi and Silver (1999a, 2001) tend to see financial and commercial capital as alternating expansions following crisis, "resumed when a new hegemonic state opened up a different developmental path, reorganizing the system so as to solve the problems and contradictions encountered along the path opened up by its predecessor."

## 11.8 Foundations for policy

The evidence reviewed here supports the argument that commercial markets with multiple buyers and sellers (multiparty multi-cohesive exchange networks) support fair pricing in equilibrium markets. They are massively different from the military-backed market conquests that, since 1450, have imposed price differentials in favor of dominant polity price-setting in regions conquered by Europeans. "Western" trade practices established unfair pricing outside the multi-cohesive network cores of the dominant economic sectors. That a core–periphery network of trade structure came to sustain unfair pricing (what Marxists call unequal exchange, including structural inequality) seems indisputable.

Further, because of the historical connection between state-policy conquest trade, the financial capital needed for European conquests, and the dependence of the early European-constructed world system (starting *circa* 1450) on bullion (soon, from the Americas) in trade for Asian luxury goods, finance capital has largely dominated the state policies of leading European economic sectors. When these sectors collapse, in their competition with other leading states, there are periods of crisis and uncertainty in which commercial markets, organized on a basis that differs from that of those sectors in crisis, and on a bottom-up basis, are able to establish fair-trade equilibrium pricing. Commercial markets also survive, innovate, and operate interstitially alongside mercenary leading-sector financial markets, once they are re-established.

A derivation from these observations is that those institutions that are protected during periods of leading sector dominance (what Arrighi calls hegemons), including banks, operate on a "maximize profits by any means necessary" basis, without regard to other players. This type of system, then, has been historically biased towards inequality, not only in terms of international core–periphery market exchange, within core states, but even more severely, within periphery and semi-peripheral states, cities, and regions.

Because the mercenary form of militarized capitalism in the European context was a reaction to the predicament of exchanging bullion or luxury goods in East–West trade we should not conclude that capitalism in the form of dominant financial markets or hegemonic and structural inequality, as distortions of pricing, are inexorable or the "natural" driving force of "evolutionary learning." Nonetheless this condition persisted and intensified until both India and China were brought under British rule. India was conquered after 1857 and colonized. China was subjected to the Opium trade from which it was barely freed in 1910. The historical specificity of each of these processes militates against the idea of inexorability.

*Ethical* capitalism, under a commercial rather than a predatory "Guardian" ethic of military-backed nationalism, remains an unexploited possibility that can still be experienced locally (e.g. in regulated S&L banking) if not internationally. Attempts to establish a British commonwealth of nations began in 1884 and culminated after the first world war in the Balfour Declaration (1926), formalized in 1931. Attempts were made to establish fair trade practices in spite of massive core–periphery structural inequalities, which were not removed by post-second world war independence movements that exposed further financial vulnerabilities of new nations. These vulnerabilities were and still are exploited by hegemonic economic practices, even in World Bank loans and IMF takeovers involving institutions still dominated by players in the leading economic and political sectors.

Currently, the core financial economy is so overconnected (Schweitzer *et al.*, 2009a,b) that there is hardly any possibility of full transparency without new systems of monitoring accountability that outperform the speed of e-transfers. Collusion and pilfering of funds can be introduced at trillions of places due to the vulnerabilities of contemporary lapses in transparency. Oversight or regulation is currently miniscule compared to the volume of transactions of vast unregulated networks of financial transmission. Failure to advocate and achieve full transparency and selective regulation produces an economy of structural criminality with potential whistle-blowers paid maximally not to tell what they know. Price collusion as to departures from fair trade can be detected with econometric methods (e.g. White, 1998, 2004) but these methods are not used currently outside of lawsuits on behalf of victims of price collusion.

The generation of hyper-inequality operates on top of the multiple problems of fair markets put out of joint by financial institutions and by state-driven trade policies. Turchin's study of structural demographics shows that overpopulation cycles that generate scarcity accelerate the concentration of capital ownership and under-payment of returns to labor. This creates incentives to manage policy to further increase inequality in managing state surpluses and civil benefits. Social conflict and economic collapse are then amplified well beyond the capital replacement costs of repairing infrastructure that are at the root of Kondratieff cycles. These replacement costs involve periods of borrowing and investing new capital directed towards innovation as well as infrastructure renewal. In the political–financial quagmires of the 2010s, awareness of national level replacement costs seems to have disappeared in public discourse. New understandings of the shapes of Kondratieff waves and phases are needed in terms of what to do about economic pricing distortions due to structural and hegemonic inequality, rather than the realpolitik of the Modelski model of power-politics, fixed on the state-policy levels of leading-sector competitors for hegemonic dominance of the international economy, as though these were the only lessons in "global learning."

## 11.9 Combining causal analysis and historical recurrence

Metastable and thus irregular cyclical behavior, however, is a normal and potentially self-regulating or beneficial condition of many complex systems (Lawless *et al.*, 2010). The empirical historical data discussed in this chapter (and in related sociological analyses of White *et al.*, 2011) show that Turchin's models of secular cycles of large-scale agrarian systems, once they combine with monetization, produce exceptionally destructive tendencies. These include: (1) bid/ask market pricing inequality amplified by the dichotomy between property owners and workers in periods of Malthusian scarcity, that produce hyper-inequality and consequent adverse effects; (2) economic downturns that add to the severity of socially violent economic crashes or depressions, followed by unusually innovative periods that go beyond population recovery to cause the impoverishment that encourages intra-family self-help and exponential population growth – until another crisis of cyclical growth occurs.

In contrast, without bid/ask amplification of inequality to amplify socially conflictual hyper-inequality (and without hyper-aggressive state-driven trade policies) it is conceivable that upturns in cyclical recoveries would generate less population growth with the effect of "demographic revolutions" wherein economic growth generates incentives (e.g. costs of children's education come to vastly outweigh benefits of children's labor) that encourage smaller family size. The social disruptions and conflicts that come out of Turchin's agrarian crisis dynamics, however, – given the context of European hegemonic trade dominance after 1450 – typically produce larger families and, overall, exponential global population growth.

Mathematical models of policy incentives do not take into account runaway systems that represent defects in recurrent cyclical behavior, or schismogenesis. The impulse of "mechanism design" might be in the right direction,[5] but may be too piecemeal for what are larger systemic or structural problems in national and international economic policies and structures.

## 11.10 Policy conclusions

The most comprehensive areas of sustainable commercial trade define a different concept of cost and price by evaluating the cost of reproducing goods, including the ecological components that are diminished in the process of collection,

---

[5] For economic "mechanism design" examples see the IMBS conference, described in terms of: "A need to understand how a system of interaction, rewards, etc. can be created in order to accomplish a desired outcome is a goal that is shared by several disciplines, ranging from political science, economics, engineering, and computer science: this is one of the objectives of mechanism design. Closely related is the need to understand how biological and other systems adapt to new circumstances: this is the area of adaptive systems."

depletion of raw material, production, and recycling. Not only, for example, is soil depleted by 300% overuse of nitrate fertilizers after germination (Agency for Toxic Substances and Disease Registry, 2007), but runoffs destroy the viability of waterways, aquifers, estuaries, lakes, and oceans. Industries that support nitrate commodity chains and use are opposed to sustainable agriculture even though poisonous water runoff kills fish and marine flora and biota worth trillions. It is evident that nitrate use and pricing, given the scale of ecological destruction, are wildly out of equilibrium. Fixing nitrate contamination, which destroys ecologies and the complex efficiencies of biodiversity, would also entail new kinds of food production that recycled existing waste differently and different kinds of foods being produced. Subsidies would need to change, and possibly how new bio-efficient foods are packaged and processed by industrial machinery, and how genetically modified seed stocks are supervised. If "costs" of sustainability are seen by corporations as eliminating sales and jobs in "environmental industries," activity could shift to creating new work to fix new problems and to recover value in ecological resources currently being destroyed. But who represents these balancing interests? For Costanza *et al.* (1997): "Because ecosystem services are not fully 'captured' in commercial markets or adequately quantified in terms comparable with economic services and manufactured capital, they are often given too little weight in policy decisions. This neglect may ultimately compromise the sustainability of humans in the biosphere." In the current change in concepts of cost and prices, "ecosystem services" are being factored where ecological costs are present not only as "sellers" but as brokers for positive returns of ecological value.

Recognizing costs in truer forms of pricing are opportunities in themselves and for new sustainable industries. For Sassen (2009), going further, "Cities are a type of socio-ecological system that has an expanding range of articulations with nature's ecologies. Today, most of these articulations produce environmental damage. [She] examines how we can begin to use these articulations to produce positive outcomes – outcomes that allow cities to contribute to environmental sustainability. The complex systemic and multi-scalar capacities of cities are a massive potential for a broad range of positive articulations with nature's ecologies."

Multi-agent and multi-organization (multiparty) policies are discussed in many venues, the European Union, United Nations, and World Trade Organization, among others. Ostensibly they are aimed at mitigating inequalities of the terms of exchange, and should be helping to shape the resultant modulation of cyclical economic fluctuations. The creation of a more level playing field for multiple agents competing in economic exchange could mitigate the emergence of hyper-inequality that may otherwise occur in key population-growth phases. The problem is exacerbated, however, by the fact that the WTO is effectively run by four hegemonic interests: the USA., Britain, France, and Germany. Even within those countries, the WTO representatives do not represent the interests of broadly different sectors, but the sectors with major hegemonic interests.

Modulations from the WTO and other international institutions are needed that serve, instead, to lessen the attendant dangers of regional and global population growth that produce recurrent Malthusian crises. Hyper-inequalities or moderating influences may operate in various ways in global crises caused by scarcity relative to population. It is in these periods of economic and social disequilibrium where ownership often comes to entail massive economic advantage in a context of scarce-resource overpopulation diminishing benefits not only to workers but to the ecologies on which they depend. Reforms in "democratizing" the WTO and other international institutions supposed to serve common interests are consistent with the aims of this chapter. That is, "democratizing" should assert the need and potential for ethical systems commercial codes to govern economic and globalization policy. The aim should be to equalize disparities in pricing and how this could lead to benefits in issues of human survival. Oka and Chapurukha (2008) and Oka *et al.* (2009), for example, document how commercial codes of exchange promoted resilience to "Famines, Maritime Warfare and Imperial Stability in the Growing Indian Ocean economy" *c.* 1500–1700 CE. Their ethnographic-historical studies represent local level operation of commercial codes in Africa and the Indian Ocean, which continue in subdominant trade today. Ethical systems commercial codes are operative worldwide throughout recent centuries, but they operate at the margins rather than in the leading economic sectors.

## 11.11 Policy alternatives for a structural-demographic complex

One set of new global policy objectives should engage in understanding the negative consequences of overpopulation cycles (Malthusian oscillations) leading to hyper-inequality and to global overpopulation. Considerations might include (1) regulation of trade through the commercial ethic and egalitarian policy institutions equipped with regulatory rules supported by, but not primarily dependent on ratification by states, (2) state policy support for the implementation of rules for leveling economic exchange and forestalling such deep oscillations of inequality as to create exploitation and civil distress (e.g. think: Wisconsin and its governor's desire to criminalize the right of unions to negotiate), (3) state and international incentives for regulation of human population that also help dampen oscillations of population growth and hyper-inequality in large regional exchange economies, and (4) international laws and regulations that enact legal liabilities on corporations so that, as with the freedom of individuals to interact in ways that generate beneficial feedback processes, they are guided toward the ethic of commerce and free exchange, rather than the Guardian ethic of both individual and corporate lying and loyalty at the expense of others as "opponents" rather than exchange partners.

## 11.12 The need to replace the Eurocentric "Guardian" model of evolutionary learning with a geocentric transparency model

A second set of new global policy objectives should be to sustain in a positive way the challenges not of, but in recognition of, the facts of Modelskian view, which I take as accurate, in our current new millennium. The notion that the "natural" learning curve of human evolution is military-backed financial capital must be challenged. The USA has nearly lost its claim to leadership as the leading economic and political power, with three unsuccessful wars (Korean, Afghan, and Iraqi) and failed regulatory economic policies. The Afghan war was a policy blunder, especially in the light of reports from anthropologists at the time that the Pashtun leadership was reconsidering their "guest" policy towards the Al Queda leadership prior to the American attack, and the historically evident fact that a war against Afghan tribal and clan warriors and incommensurate codes governing social norms was unwinnable. At the moment that I am writing this, with the people's revolution of the ethnically unified central nation of Egypt in the Middle East, we have a step of democratization that could act to level the economic field of hegemonic pricing and monopolization of militarized power.

## 11.13 Transparent ports of trade as "evolutionary learning" on how we get to policy alternatives for fair trade and sustainability

Two forms of commercial code exchange can be imagined. One echoes West's view, "It's the freedom of the city that keeps it alive," i.e. transactions within a city (or town) allow reputational knowledge of merchants and fair prices, at least as judged locally. This does not apply, however, to international or Internet trade, although the Internet allows customer ratings of merchants and comparison sites for prices. Customer ratings can, however, be rigged, by merchants themselves making bogus ratings, preventing price-quality ratios from being judged fairly.

What is the possibility for ports of trade, like the Champagne Fairs, that provide complete transparency? That is, reputations of merchants, recognition of conflicting interests, subsidies derived from motivated contributions, fully disclosed fees and pricing mechanisms, enforcement of regulations against insider dealing (as when arbitrage sales hand collateral of short sellers over to banks that return interest on the stocks to the arbitrageur), knowledge of provenience of goods and the cost markup of price at each stage in its transfer, etc. At this moment arbitrage in the USA is supposed to be "neutral" in removing non-equilibrium pricing, but is totally non-transparent and managed by nine individuals representing different banks. A transparent system must prevent the possibility of collusion by providing full information that would detect the sources of bias, and investigate such possibilities post-hoc. It must have in force transparent and verifiable information that would allow buyers or customers to avoid unfair pricing, monopsony, monopoly (more generally: oligopoly and oligopsony), and a system of regulations and incentives or punishments for unfair or deceptive behavior, including disbarment from the market. In principle, a free and fully transparent market is self-equilibrating, but only if the status of participants in and of itself does not affect prices, which even Aristotle observed as a causal principle of economics. In principle, price of entry can be free but is only meaningful provided that disbarment or posting of information about unethical behavior is insured.

National economic policies need to turn from viewing warfare as a putting out of political fires and eliminating threats to overseas economic interests – which include illicit policy support for military–industrial revolving doors between its leading industries, including international weapons sales. It is necessary to comprehend the heightened effects of cyclical (and currently growing) inequalities in regions all over the world, as a result of national policy priorities toward hegemony over other national economies on a competitive military-backed priority. Instead, reform of WTO and other institutions must aim at making peaceful alliances and a level playing field – without chokepoints, monopolies/monopsonies, or national and international oligopolies designed to uphold purely national interests. In principle, transparent trade associations among nations could mirror transparencies in nonstate trade associations. The more fully international each of these "transparent trade associations," state or nonstate, the greater potential for a level economic playing field.

## 11.14 Conclusion

Without intending to in advance, but attentive to the history of "globalization," I believe I have documented a case for concluding that it is the militarization of economic growth (1450–present) in leading economies that sustains global impoverishment. This in turn erodes ecologies and incentivizes large families as a means of survival. Pitting population growth against fixed, slowly growing or declining resources causes Malthusian crises that generate hyper-inequality in a vicious cycle, which at every level entails destructively unfair pricing and a tendency for commercial ethics to degrade to those of the Guardian warfare-bases ethics. Economic growth cannot be sustained under the conditions of military-backed national trade policies of European globalization that began in 1450 and have continued to the present. Foundational change is required in the militaristic policies of leading economies. This view contravenes the Modelskian view of "global learning" naturalized as militaristic globalization.

If this view is valid in its long-view diagnostics, it entails that leading economies will do better in a global economy when they reduce their military power to achieve equilibrium with other powers, including grand alliances among democracies as against newly militarizing autocracies. Abstaining from economic advantage taken from military power, and supporting viable international institutions such as GATT or a reformed WTO can help lead to both equilibrium market prices that are fair as between nations, and to international regulatory agencies favoring fairness that provisions the future of all nations and of a growing and diversified ecology, that can support growing economic benefits for individuals and not simply national GDPs but GDPs per capita worldwide. In these circumstances, the diversification of sources of wealth in the absence of militarily backed economic inequality supports diversified labor as well as diversified rights and ownership. This reconfiguration of policy priorities can re-establish the basis for moral action over dishonesty, distrust, and localized interests aimed at the destruction of others. These reconfigured conditions can produce local and globalized demographic transitions towards preferentially smaller families and the lowering of world population, lessening the likelihoods of cyclical occurrence of regionally Malthusian crises that promote conflict, warfare, and collapse which, once completed, lead to periods of intense population growth that exceeds previous levels of population by virtue of excessive innovation. This reconfiguration can impede an arms race aimed at competition with other peoples and nations which is the current source of Modelskian "global learning," a theory that itself needs to be reconfigured in a more proactive way than a Hobbesian war of all "Guardian" interests against all others.

## 11.15  Reprise and critique

In the context of twentieth and early twenty-first century globalization, the post-2007 crisis research of economist Coyle (2011) supports the arguments of this chapter. Her chapter on fairness (and causes or consequences of extremes in inequality) stands as a recommended supplementary review of evidence for the reader. That chapter, however, does not mention fairness in pricing per se, or the role of leading sector state policy in employing military power in a myriad of ways that alter prices. Focusing only on recent times misses the long trajectory of military-backed attempts to alter pricing by European-based powers, from Portugal (1450) to present-day American power. I have no personal beef with the professional military, my father and grandfather being a Major General and a Lieutenant Colonel in the second and first world wars, respectively, but only with the misuse of military power and its consequences, and in arguing against the view that the theory of competitive pricing equilibrium represents a basis for the view of economy of the last 560 years of globalization as a level playing field.

My views on reducing the costs of warfare for civil societies and nations will be criticized by those who hold that warfare is an inevitable and immutable feature of competition. Consider as an example how in the USA, the rush to punish Al Queda failed to parse the signal that Mullah Omar and Taliban leaders were reconsidering the Pushtun doctrine of hospitality in the possibility of turning over the Al Queda leadership to Pakistan, similarly to the Sudanese offer to Madeleine Albright a decade earlier. The cost of the hothead rush to war without diplomacy and discussion has been in the tens of trillions. That kind of rush to warfare by leading powers is potentially containable. Klein's (2008) four-year research project on contemporary globalization, summarized in *The Shock Doctrine: The Rise of Disaster Capitalism* contains what has been called "nothing less than the secret history of what we call the 'Free Market.' " by Arundhati Roy, who adds "It should be compulsory reading," especially as it is playing out today (March 2011) in gubernatorial policies in Wisconsin, Michigan, and elsewhere aimed at relieving Americans of their rights as citizens.

The reader dubious of my propositions about containment of warfare might also consider how, in the current century, the concept of warfare logistics has been yoked to corporate outsourcing, again at a cost of tens of trillions. The problem here is that the agents of corporate outsourcing have a vested interest in warfare as a source of profits, which contaminates the concept that a democratic government should be run by civilians, and officials hired by governments should not be contaminated by conflicting interests in the war machines hosted by corporations rather than by professional militaries.

## 11.16 Acknowledgements

I am indebted to the Santa Fe Institute for hosting my participation in the Co-Evolution and Networks and Markets Working Group of the Santa Fe Institute, and funding an SFI Working Group on "Analyzing Complex Macrosystems: Civilizations as Dynamic Networks," to Bob Adams for suggesting that I read Peter Spufford's (2002) book, to Peter Spufford for starting down the path of studying the Medieval Renaissance and suggesting that a trade network study of economic flows between Europe, China, and India would be needed, to Peter Turchin for sharing his knowledge and images of historical dynamics, and to John Padgett for the invitation to join the Working Group on Co-Evolution of States and Markets, Santa Fe Institute, 1999–2002.

# 12

# Network science in ecology: the structure of ecological communities and the biodiversity question

ANTONIO BODINI, STEFANO ALLESINA, AND
CRISTINA BONDAVALLI

The study of networks in ecology is rapidly expanding. Although network thinking is by no means new to ecologists, cross-fertilization from other fields, ranging from computer science to sociology, has recently furthered the field significantly. Here we examine some of the applications of network science to ecology, with an emphasis on its potential to contribute to the preservation of biodiversity, an issue that has relevant social and policy implications. Two different forms in which ecological networks may appear are used: food webs and signed digraphs of dynamical systems. In the former, networks represent energy flow transfers from producers to consumers, while in the latter what is depicted is the effect that populations exert on each other.

The main objective is to enlighten as to how applying network science can contribute to some central questions concerning biodiversity, such as the identification of keystone species, the response of population to environmental perturbations, the robustness or inertia of the system to external events in the form of loss of species and links that alter population dynamics.

## 12.1 Biodiversity and the network perspective in ecology

In the last decade biodiversity loss has become of major concern (Loreau *et al.*, 2001; Ceballos and Ehrlich, 2002; Pimm *et al.*, 2006). In the face of this crisis, policies aimed at preserving biodiversity have been called for (Westman, 1990). To shape effective management strategies, a great deal of effort is required in a diversity of fields (Peuhkuri and Jokinen, 1999), prominantly, ecology. Over the course of many years ecology has developed a formidable background that can support studies on biodiversity, and tools that are commonly used for

Networks in Social Policy Problems, eds. Balázs Vedres and Marco Scotti. Published by Cambridge University Press © Cambridge University Press 2012.

this purpose are well shaped in the various domains of ecological knowledge: from community ecology (diversity indices, distribution and abundance of populations, species interactions, and dynamical stability) to population dynamics (extinction and colonization, meta-populations, population viability, speciation) and organism physiology (tolerance to stressors, individual basis for interaction mechanisms).

In the last decade, however, there has been an increasing awareness that also the structure of the community in which species are embedded, that is the vast array of interactions that occur between the species, can be relevant. Distant fields of science provide evidence that the anatomy of complex, web-like structures is the appropriate locus of explanation for functions and dynamics of systems. For example, the topology of social relationships may help in assessing the magnitude of the health risk in human societies (Cohen *et al.*, 2000b); the anatomy of trade flows between countries may shed light on the dynamics of global economic processes (Krempel and Plumper, 2003); and the properties of metabolic networks can explain the tolerance of simple organisms to environmental modification or pharmaceutical intervention (Hartwell *et al.*, 1999). It therefore become natural for ecologists to consider understanding the structure of interactions as crucial for achieving a correct grasp of ecosystems functioning. Juxtaposing in pictorial form all the interactions between species within a community one obtains entangled networks known as "food webs." Typically, the structural elements, the connections, are of a trophic nature: they represent who eats whom (and by how much) in an ecosystem. Since the pioneer investigation of the Italian zoologist Lorenzo Camerano (1880; compare Cohen, 1994) food webs have been extensively investigated (Paine, 1980; Cohen *et al.*, 1990; Pimm *et al.*, 1991), but only recently have they been regarded as targets for applying principles and concepts of network science (Proulx *et al.*, 2005; Pascual and Dunne, 2006). What nowadays makes food webs attractive to ecologists is that: (a) they can be used to define the role of species with respect to certain ecological functions (energy delivery, robustness, stability, and resilience); and (b) they play a key role in the rise and collapse of species by transmitting fluctuations of one species' abundance to the others through complex pathways of interactions. Both these features may have profound implications for biodiversity. In spite of the presently resurgent wave of interest for food web analysis in the form of network thinking, the importance of the structure of the interactions in ecological communities as a functional determinant had already been recognized in the 1970s. In his "The qualitative analysis of partially specified systems" (1974), Richard Levins stated that: *"complex (ecological) systems may be studied (qualitatively) by an examination of network properties of the whole."* The conceptual and mathematical apparatus developed by Levins, which is often known as *"loop analysis,"* may contribute towards answering questions related to points (a) and (b) above; furthermore it allows deeper insights into the dynamics of ecological communities, especially for understanding cause–effect mechanisms.

In this work, we would like to show the relevance of network thinking in ecology and, in particular, what it may contribute to the biodiversity issue. It is not our intention here to provide a review on the topic, as there are several outstanding syntheses in the literature (May, 2006; Pascual and Dunne, 2006; Bascompte, 2007); rather we will showcase our own research to describe, through some examples, the types of argument that can be dealt with by using food web analysis and loop analysis as main expressions of network applications.

## 12.2  Food web networks and the problem of secondary extinction

Albert *et al.* (2000) introduced the analogy of biological extinctions into computer science, studying how random or targeted disconnection of some servers would affect the connectedness of the Internet network (cascading disconnections). Applied to ecology this would imply that a single extinction event would eventually cascade into further extinctions because species in ecosystems "communicate" by controlling each other's growth rate through predation, competition, and mutualism in multiple reticulate connections. Ecologists dubbed this phenomenon "secondary extinction." The ultimate outcome of catastrophic, multiple secondary extinctions is the collapse of the entire food web. In this context, the interest of ecologists has concentrated on the degree to which an ecosystem may experience extinctions and still be a functioning whole. This property is called robustness, and fragility is its opposite. These issues, framed in a different methodological context, with a new vocabulary, have reoriented and magnified the long-standing question of stability of natural communities and ecosystems (May, 1973; McCann, 2000; Kondoh, 2003; Montoya *et al.*, 2006); but with the merit of reconsidering correctly the question, originally posed by MacArthur (1955) – see Ulanowicz (1997) and Allesina *et al.* (2009). Secondary extinction has been investigated in a topological perspective (Dunne *et al.*, 2002a,b, 2004; Melián and Bascompte, 2002; Montoya and Solé, 2002; Ebenman and Jonsson, 2005). Particular attention has been paid to species that by means of multiple interactions behave like hubs in the network. Once removed, the food web network they belong to would quickly fragment into isolated clusters with the extinction of many species, and the system as a functioning whole would collapse. These hub species have been indicated as keystone species. The keystone species has become a core concept in the ecology and conservation literature since it was first introduced by Paine (1969). Keystones are those species whose removal (extinction) would cause a dramatic restructuring of the community (Estes and Palmisano, 1974), usually accompanied by species loss. In spite of the many empirical observations of this phenomenon, corroborated by *in situ* experiment (Navarrete and Menge, 1996), the mechanisms that make a species keystone are only partially understood. Food web analysis may add to the scientific debate on keystones by addressing the question from the network point of view. Two attributes seem to play a decisive role in this respect: the position one species occupies in the network (Jordán, 2009), and its connectivity, the fraction of realized interactions (Dunne *et al.*, 2002a,b).

Robustness, however, is not simply determined by topological attributes; structure is always associated with function and one crucial question is how food web architecture governs energy distribution and availability to the component species. In this framework, secondary extinctions can be reinterpreted as follows: whenever a species that is eaten by another disappears and the consumer has no alternative resource to exploit, it will also become extinct. This mechanism directly address to robustness sensu MacArthur (1955): a higher number of links would enhance stability because of the large number of pathways that can be used for the transfer of energy. The multiplicity of connections may also buffer the consequences of food web alterations, as it may provide alternative routes for species to get their requisite medium when the main provider of food vanishes. To unveil the circumstances in which food web structure amplifies or buffers the effects of species loss is of fundamental importance. In this context, keystone species are only those that are strategically positioned within the flow of energy in the food webs so that many other species depend on them for their requirements. The analysis of the food webs in search of these bottlenecks can be conducted using a graph theory tool called dominator tree (Allesina and Bodini, 2004).

### 12.2.1 The dominator tree

Matter and energy travel in food webs from producers to consumers following complex pathways formed by link juxtaposition. Certain routes are obligatory when there are no alternative paths for energy to flow from one species to another. A dominator tree is the topological construct that groups the whole suite of these obligatory routes. Any species in these chains is said to dominate the followings because they necessarily depend on it to satisfy their own energy requirement. This is why these topologies are called dominator trees.

By these structures one can easily identify which nodes are likely to cause the greatest impact if removed. Consider the hypothetical food web depicted in Figure 12.1 (graph on the left). All its nodes are connected directly or indirectly to a virtual node called Root. In it we collapse the external environment as the ultimate source of energy for the entire ecosystem.

One of the fundamental theorems of dominator trees (Lengauer and Tarjan, 1979) states that "*every node of a graph except the Root has a unique immediate (i.e., closest) dominator.*" Accordingly, we can build a dominator tree by connecting each node to its immediate dominator. This procedure can be simplified as follows.

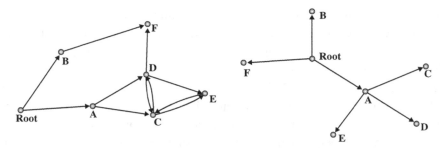

Figure 12.1. Simple hypothetical network (left) and its corresponding dominator tree (right). The removal of node *A* would cause the disconnection of nodes *C*, *D*, *E* while the removal of any other node different from the *Root* would cause no secondary extinction (Allesina *et al.*, 2006).

Species *E* receives medium from species *C* and *D*, both of which, in turn, depend on *A*. The extinction of either *C* or *D* does not put *E* at risk of survival because at least one pathway remains at its disposal. In fact this latter species receives its medium by means of two pathways that are *Root* → *A* → *D* → *E* and *Root* → *A* → *C* → *E*. On the contrary, extinction of *A* inevitably drives *E* (and *C* and *D* as well) to extinction. The energy that reaches *E* passes through *A*, which is the unique species necessary for *E* to survive. Accordingly, the dominator tree traces a direct connection from *A* to *E*. Given that *A* is essential for *C* and *D* as well, the dominator tree depicts a direct connection from *A* to each and all of these three nodes. Because the dominator tree illustrates through its direct links only the dominance relations, there are no connections between *C*, *D*, and *E* because none of them is essential for the survival of any of the others. We state that given a network *G* with *N* nodes (*N* − 1 species and the root) connected by *L* edges associated with weight *W*, a node *A* is a dominator of node *B* (*A* = *dom*(*B*)) if and only if every path from *R* (*Root* of the network) to *B* contains *A*.

An interesting case, particularly useful for understanding the method of dominator trees is that of the Grassland ecosystem (Martinez *et al.*, 1999). This comprises a relatively small number of nodes and, at the same time, a sufficient number of hubs so that the problem of secondary extinction can be appreciated visually. Unfolding the food web of this ecosystem by identifying dominance relationships yielded the graph of Figure 12.2 which clarifies dominance relations between nodes.

A few nodes (e.g., 1, 2, 3, 4, ..., 51, 53, 61, see figure caption for correspondence between numbers and species) are dominated by the Root only: these species cannot be extinguished by removing other taxa. Nodes 4, 38, 51, 53, 61 do not dominate other species. The rest of the nodes linked directly to the Root initiate a branch. Removing these nodes would make a variable number of species disappear. For example, removing node 3 – *Elymus repens* – would cascade up to nodes 13, 14,

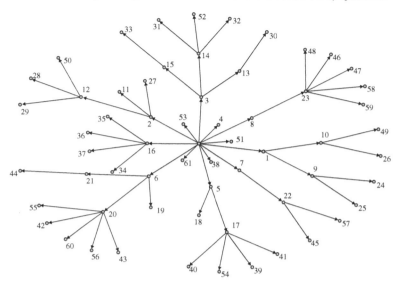

Figure 12.2. Dominator tree for the Grassland ecosystem (Martinez *et al.*, 1999), which comprises grass species and their associated endophytic insects in Great Britain. Correspondence between numbers and species is as follows: R: *root*; 1: *Festuca rubra*; 2: *Alopecurus pratensis*; 3: *Elymus repens*; 4: *Ammophila arenaria*; 5: *Calamagrostis epigejos*; 6: *Deschampsia cespitosa*; 7: *Dactylis glomerata*; 8: *Phalaris arundinaceae*; 9: *Tetramesa brevicomis*; 10: *Tetramesa brevicollis*; 11: *Ahtola atra*; 12: *Tetramesa angustipennis*; 13: *Tetramesa linearis*; 14: *Tetramesa hyalipennis*; 15: *Tetramesa comuta*; 16: *Tetramesa eximia*; 17: *Tetramesa cala-magrostidis*; 18: *Eurytoma sp.*; 19: *Eurytoma sp.*; 20: *Tetramesa petiolata*; 21: *Tetramesa airae*; 22: *Tetramesa longula*; 23: *Tetramesa longicomis*; 24: *Bracon sp. + Sycophila sp.*; 25: *Homoporus sp. + Eurytoma sp.*; 26: *Eurytoma sp.*; 27: *Chlorocytus pulchripes*; 28: *Sycophila sp. + Homoporus febriculosus + Endro-mopoda sp.*; 29: *Eurytoma tapio*; 30: *Sycophila sp. + Pediobius sp. + Eurytoma flavimana + Eurytoma sp. + Homoporus sp.*; 31: *Bracon erythrostictus*; 32: *Eury-toma roseni*; 33: *Chlorocytus agropyri + Pediobius alaspharus*; 34: *Eurytoma sp.*; 35: *Homoporus sp. + Sycophila sp. + Eurytoma danuvica*; 36: *Homoporus ful-viventris*; 37: *Chlorocytus harmolitae*; 38: *Syntomaspis baudysi*; 39: *Bracon sp. + Homoporus luniger*; 40: *Endromopoda sp.*; 41: *Eurytoma pollux*; 42: *Eurytoma appendigaster*; 43: *Homoporus sp.*; 44: *Pediobius sp. + Chlorocytus sp. + Eury-toma castor*; 45: *Eurytoma erdoesi*; 46: *Bracon sp.*; 47: *Eurytoma phalaridis*; 48: *Endromopoda sp.*; 49: *Pediobius festucae*; 50: *Pediobius eubius*; 51: *Eupelmus atropurpureus*; 52: *Endromopoda sp.*; 53: *Macroneura vesicularis*; 54: *Pediobius calamagrostidis*; 55: *Pediobius deschampiae*; 56: *Endromopoda sp.*; 57: *Pediobius dactylicola*; 58: *Chlorocytus phalaridis*; 59: *Pediobius phalaridis*; 60: *Chlorocytus deschampiae*; 61: *Mesopolobus graminum*. See Allesina and Bodini (2004).

and 15, which would vanish. In turn, these secondary extinctions would propagate to nodes 30, 31, 32, 33, and 52. This example shows how dominator trees can be utilized to predict the effects of species removals.

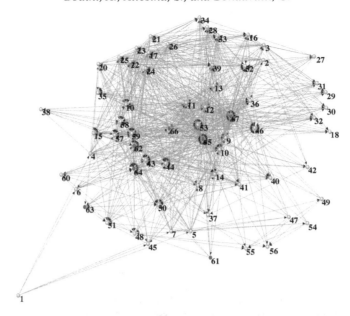

Figure 12.3. Gramminoid Marshes ecosystem food web (this page), and corresponding dominator tree (following page). Correspondence between nodes and species/group of species/pools: 1: *Root*; 2: *Living sediments*; 3: *Living POC*; 4: *Periphyton*; 5: *Macrophytes*; 6: *Utricularia*; 7: *Floating veg.*; 8: *Apple snail*; 9: *Freshwater prawn*; 10: *Crayfish*; 11: *Mesoinverts*; 12: *Other macroinverts*; 13: *Large aquatic insects*; 14: *Terrestrial invert.*; 15: *Fishing spider*; 16: *Gar*; 17: *Shiners and minnows*; 18: *Chubsuckers*; 19: *Catfish*; 20: *Flagfish*; 21: *Topminnows*; 22: *Bluefin killifish*; 23: *Killifishes*; 24: *Mosquitofishes*; 25: *Poecilids*; 26: *Pigmy sunfish*; 27: *Bluespotted sunfish*; 28: *Warmouth*; 29: *Dollar sunfish*; 30: *Redear sunfish*; 31: *Spotted sunfish*; 32: *Other centrarchids*; 33: *Largemouth bass*; 34: *Cichlids*; 35: *Other large fishes*; 36: *Other small Fishes*; 37: *Salamanders*; 38: *Salamander larvae*; 39: *Large frogs*; 40: *Medium frogs*; 41: *Small frogs*; 42: *Tadpoles*; 43: *Turtles*; 44: *Snakes*; 45: *Lizards*; 46: *Alligators*; 47: *Muskrats*; 48: *Rats and mice*; 49: *Rabbits*; 50: *Raccoons*; 51: *Opossum*; 52: *Otter*; 53: *Mink*; 54: *White tail deer*; 55: *Bobcat*; 56: *Panthers*; 57: *Grebes*; 58: *Bitterns*; 59: *Ducks*; 60: *Snailkites*; 61: *Nighthawks*; 62: *Gruiformes*; 63: *Cape Sable seaside sparrow*; 64: *Passerines*; 65: *Sediment carbon*; 66: *Labile detritus*; 67: *Refractory detritus*. For further information on species please refer to the ATLSS website (http://www.cbl.umces.edu/~atlss/swgrasøør.html). This network contains 67 nodes and 798 links (see Allesina *et al.*, 2006).

### 12.2.2 Potential and limitation of the dominator tree

Dominator trees are elegant, highly informative structures that allow identification in advance of which nodes (species) are likely to cause the greatest impact on the web if removed. This leads immediately to practical questions. It is common wisdom that keystone species have large effects on other species in the community.

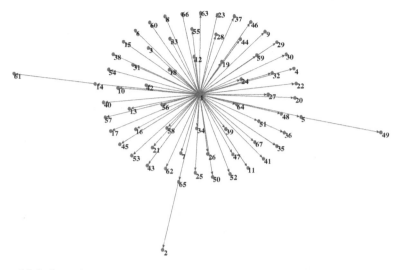

Figure 12.3  *(cont.)*

Accordingly, attention should be devoted to the conservation of these taxa, and methods suitable for their a-priori identification have been called for. Dominator trees have much to offer because they simplify food webs so that keystones may be easily identified. Any model should be tested for its forecasting efficacy. To assess the predictive potential of the dominator tree, Bodini *et al.* (2009) compared dominator predictions with the observed effects that followed the collapse of the capelin (*Mallotus villosus*) in the Barents Sea ecosystem. Results highlighted that although some correspondence between prediction and observation was found, the dominator tree was only partially satisfactory. In practice, of the three species (the kittiwake, *Rissa tridactyla*; the Brunnich's guillemot, *Uria lomvia*; the common guillemot, *Uria aalge*) that the model showed to be dominated by the capelin, and thus prone to extinction, only the latter really collapsed in the Barents Sea, and one that went locally extinct (the harp seal, *Phoca groenlandica*) was not predicted to extinction by the model. To explain this incongruity between prediction and observation it is necessary to consider that the dominator tree is a pure qualitative architecture. So when a species feeds on multiple preys, none of them dominates it. However, if one prey disproportionately contributes to the predator's energy intake, its extinction would impact the predator, no matter how many other prey species remain. Once the main prey becomes extinct, the energy provided by the remaining prey may not be sufficient to sustain the predator. This is what might have occurred to the harp seal, which underwent local extinction.[1] Link quantification in food

---

[1] This species underwent mass emigration in search of food, which for the purpose of the analysis corresponds to local extinction (Bodini *et al.*, 2009).

webs may improve the reliability of the dominator tree. When the magnitude of interactions is included in the analysis, the structure of the dominator tree changes, reshaping dominance relationships. If we assume that only links that are strong enough are essential to sustain the species, the dominator tree would come out from a parsimonious version of the food web, that retains only essential links. This requires one to know which links are important and which are not, a really difficult achievement, especially for large food webs. To provide a tentative scenario of what would happen if the relative importance of links were known, one can set up a threshold of magnitude below which links are considered non-essential for species survival and are thus discarded from the web (Allesina *et al.*, 2006). Figures 12.3 and 12.4 show the outcome of this exercise. In Figure 12.3 the Gramminoid marsh ecosystem of Florida (Ulanowicz *et al.*, 2000; Allesina *et al.*, 2006) is represented by its food web and the related dominator tree. Figure 12.4 shows what happens when only links accounting for at least 25% of each species' diet are retained as essential for species survival.

The original food web produces a dominator tree that is rather uninformative. All nodes but a few are connected directly to the root, as the vast majority of the species possess several alternative pathways to get their requisite medium; thus, few dominations emerge. This is clearly an artifact of the presence of weak links that are not weighted in the qualitative dominator tree. Once links that account for less than 25% of each node's diet are removed, only 89 links of the original 798 remain, and the dominator tree becomes more informative: forecasting key nodes for energy delivery is now easier. Although quantification effort may substantially refine dominator analysis, there are other limitations. Firstly, dominator trees frame the problem of keystone species in a bottom-up perspective. Cascading extinction also certainly occurs in a top-down direction, when predator's extinction leaves its prey without a flow of regulatory effects, but this does not necessarily imply extinction of the prey. This suggests that dominator trees account for the minimum damage that can occur de facto, that is to say a predator with no prey will certainly become extinct. Secondly, the mechanism of prey switching is not accounted for: when a predator remains without its preferred food it may switch to other resources to avoid extinction. Predator switching could easily be included by considering a food web of potential connections (i.e. including all potential prey, and not only the observed ones), and examining these "potential food webs" for secondary extinctions. Finally, another limitation of the model is that it works well when the initial extinction regards one and only one node. When the initial loss is targeted to more than one node, cascade extinctions cannot be predicted and a generalized dominator model would better account for these situations.

The concept of dominator trees, although widely used in other fields of science, has only recently been introduced into food web analysis. Its potential in the field of biodiversity conservation is promising: it could help to identify those species that are crucial for energy delivery in the community. These functionally important species should be the target of protection effort, as their conservation would prevent massive biodiversity losses.

## 12.3 Network structure and species response to disturbance: implications for biodiversity

Population decline is caused by many forms of stress due to human exploitation, invasion of alien species, pollution in its various forms, habitat fragmentation. Stressors affect the growth rate of species (i.e. by increasing their mortality), with consequences on their abundances; the problem with natural communities is that multiple interactions may distribute the effects of disturbance beyond the target of the perturbation. Unveiling the mechanisms by which an impact affecting one species percolates to other populations is essential for understanding whole community response to disturbance and consequences for biodiversity. If it is straightforward to think of a network as an appropriate tool to grasp visually the complex patterns of interactions that make up an ecological community, such representation must also offer the opportunity to understand how fluctuations in one species abundance can be transmitted to others in such a way that the observed rise and collapse of populations can be explained. That is, the analysis of the network could predict dynamics and enlighten the relationship between structural and dynamical features, a task made difficult by the presence of pervasive feedbacks. Very often surprises arise in ecosystem management (Holling, 1986) because the usual linear thinking is confounded by feedbacks (Levins, 1995). The key to grasping this complex causality is in the ability to plot the important relations in a community and understand how they interrelate to cause changes. Qualitative network analysis shows the most promise in achieving this goal.

### 12.3.1 Qualitative network analysis

Qualitative network analysis, also known as loop analysis, (Levins, 1974; Puccia and Levins, 1985) requires that interactions between species be described in their dynamical consequences, that is links must depict the effects of species populations on each other's growth rate. While in food webs, predator and prey are linked by an arrow representing the flow of energy between the two, in loop analysis links are of two types: positive ($\rightarrow$) from prey to predator and negative ($-\circ$) for the reverse. Combinations of links form pathways through which variations in any one species population propagate to the others. Pathways may connect distant species and thus effects may travel long distances in the network. The final response of one

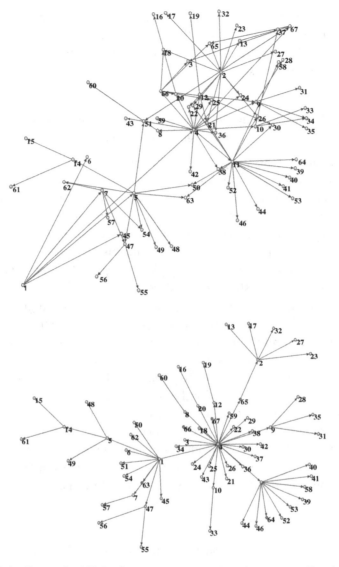

Figure 12.4. Gramminoid Marshes ecosystem (top), and corresponding dominator tree after removal of links that account for less than 25% of each species' diet (bottom). Links remaining in the food web are 89 (Allesina *et al.*, 2006).

species does not only depend on the combination of links that form the pathway that connects it to the perturbed component. Rather, it also depends on the rest of the system that can ideally be isolated from the pathway: it may create a feedback that modulates the response to the alteration carried by the pathway. This feedback is called complementary feedback. It is produced by the loop (circuit) that connects all the variables in the complementary subsystem. Finally, there is the overall feedback, produced by the circuits that connect all the species in the model. It is a measure of the system inertia: the more intense this feedback, the less pronounced is the response of the species to alterations. These concepts summarize the algorithm that is employed to predict direction of change in the abundance of species; it is condensed in the following formula:

$$\frac{\delta x_j}{\delta c} = \frac{\sum \left[\frac{\delta f_i}{\delta c}\right] \times \left[p_{ji}^k\right] \times \left[F_{n-k}^{(comp)}\right]}{F_n}, \tag{12.1}$$

where $c$ is the changing parameter (such as mortality or predation rate); $\left[\frac{\delta x_j}{\delta c}\right]$ stands for variation expected in the level of variable $x_j$;[2] $\left[\frac{\delta f_i}{\delta c}\right]$ designates whether the growth rate of the $i$th variable is increasing $(+)$ or decreasing $(-)$ because of the parameter change (positive and negative input, respectively in loop analysis language); $\left[p_{ji}^k\right]$ is the sign of the pathway connecting the variable that undergoes parameter change, $x_i$, with that whose equilibrium value is being calculated, $x_j$. Multiplying the signs of the links forming the pathway one obtains the sign of the pathway. The term $\left[F_{n-k}^{(comp)}\right]$ is the complementary feedback, while $F_n$ indicates the overall feedback of the system. The signs of both complementary and overall feedbacks are obtained in a non-simple way starting from the product of the signs of the links that form their circuits (for details see Puccia and Levins, 1985). Figure 12.5 provides the correspondence between factors in the formula and elements of a generic three-variable graph. Because only the signs of the interactions matter in this analysis, predictions are given as directions of change for biomass or number of individuals. So a species can be predicted as increasing $(+)$, decreasing $(-)$, or not changing $(0)$.

Responses of population abundance or biomass are usually arranged in tables of predictions. The entries in the table denote variations expected in all the column variables when a positive parameter change, that is an increased growth rate, affects raw variables.[3] The table of predictions is produced by altering with a positive input each variable in turn (row variable) and assessing the effect on the other variables. Predicting directions of change for species populations is what qualitative network analysis is all about. It can serve different purposes in the biodiversity context, as discussed in the next section.

---

[2] According to the theoretical apparatus of qualitative network analysis predictions regarding the equilibrium level of a variable; Bodini (1998, 2000) however has offered evidence that they apply to average values of abundance or biomass as well.

[3] Predictions for negative parameter changes may be easily obtained from the table by simply inverting the sign. For details see Puccia and Levins (1985).

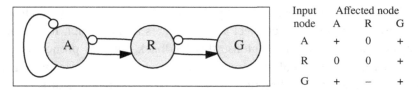

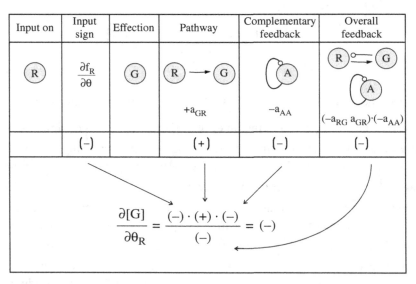

Figure 12.5. Signed digraph of a three-node network and its table of predictions. Correspondence between the formula and the structural elements of the graph that enter into it is shown, while predicting the direction of change in the abundance of species G as due to positive input on species R (R increases its rate of change); see Bodini *et al.* (2007).

### 12.3.2 The effect of variability on biodiversity: system response, the role of single species and management implications

In some cases population fluctuations are symptoms of a species' collapse. It is therefore important to have a perception of the resistance that populations may show to both external (anthropogenic) and internal (genetic, physiological) variability. Figure 12.6 shows two generic networks representing different water ecosystems.[4] They have all the components in common (see figure caption) but model 12.6b has a top predator (C) not present in model 12.6a. Each model is accompanied by its table of predictions.

---

[4] These models are very general and are by no means intended to represent specific case studies. They may describe the core set of interactions in a pelagic marine ecosystem as well as the network of a pelagic lacustrine environment (Bodini, 2000). The purpose here is not to perform predictive analysis of specific impacts, but to show the types of argument that qualitative network analysis can afford.

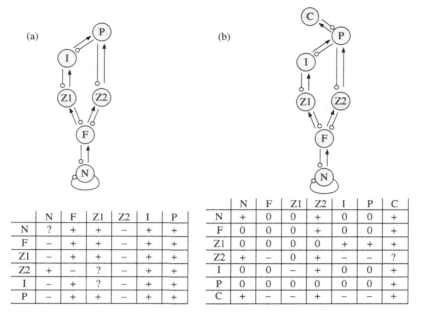

|   | N | F | Z1 | Z2 | I | P |
|---|---|---|----|----|---|---|
| N | ? | + | + | − | + | + |
| F | − | + | + | − | + | + |
| Z1 | − | + | + | − | + | + |
| Z2 | + | − | ? | − | + | + |
| I | − | + | ? | − | + | + |
| P | − | + | + | − | + | + |

|   | N | F | Z1 | Z2 | I | P | C |
|---|---|---|----|----|---|---|---|
| N | + | 0 | 0 | + | 0 | 0 | + |
| F | 0 | 0 | 0 | + | 0 | 0 | + |
| Z1 | 0 | 0 | 0 | 0 | + | + | + |
| Z2 | + | − | 0 | + | − | − | ? |
| I | 0 | 0 | − | + | 0 | 0 | + |
| P | 0 | 0 | 0 | 0 | 0 | 0 | + |
| C | + | − | − | + | − | − | + |

Figure 12.6. Networks depicting two lake ecosystems without (left) and with (right) top predator (C). Below each model is its table of predictions. Key: N, inorganic nutrients; F, phytoplankton; Z1 small-bodied zooplankton; Z2, large-bodied zooplankton; I, invertebrate predator; P, planktivorous fish; C, predatory fish; see Bodini *et al.* (2007).

Most entries in the table of Figure 12.6b are zero, while in the table of Figure 12.6a all of the coefficients are nonzero.[5] Given that a null value stands for no change expected in the abundance of the column variable, we can conclude that the system of model 12.6b shows great resistance to parameter changes, while model 12.6a is not resistant at all: all populations are expected to respond to changes entering anywhere in the system. Populations are protected against environmental fluctuations in model 12.6b but not in the system without a top carnivore (model 12.6a). Because the networks are exactly the same, except for the presence of species C in model 12.6b, we can say that this latter component is responsible for the inertia of the system. Its particular position in the network does not allow the formation of many of the complementary feedbacks that are needed to render various pathways active in transmitting the effects of perturbations. Specific ecological features of species C do not matter in this respect: what matters uniquely is its position within the network. If an additional component were to enter this system as a predatory variable on C (i.e. a predator or human exploitation), most of the null complementary

---

[5] Question marks are ambiguous predictions in the sense that the network structure produces pathways with opposite sign and nothing can be said about the new level of that variable in the absence of quantitative specifications of links. For details see Puccia and Levins (1985).

feedbacks would become nonzero, and the system would become more sensitive to environmental variability (zeros in the table would be substituted by signs). In the table of Figure 12.6b, the row and column of species C have no zeros. Changes entering the system through this node would spread their effect to the whole community, and changes entering any other variable certainly will be absorbed by C, changing its abundance. Accordingly, C can be considered as a keystone species: it protects the system from the effects of variability but at the same time it renders the system vulnerable to its own changes. Particular care should thus be given to the protection of species that are uniquely positioned in the network of interactions, like C in this example. Qualitative network analysis shows that being a keystone depends on the position a species occupies in the network with respect to the whole structure, so that it may affect the composition and the value of pathways and related complementary feedbacks that are responsible for the propagation of effects through the community.

Networks challenge our perception of cause and effect relationships. It is well known that a diffuse menace to biodiversity is eutrophication (Jenkins, 2003), that is, algal blooms caused by an excess input of inorganic nutrients into water ecosystems. To avoid the disruption of ecosystem mechanisms with fatal consequences on biodiversity, early signs of eutrophication should be detected (Bondavalli *et al.*, 2006). The most obvious surveillance strategy for early warning detection is monitoring algae abundance, but its efficacy may be defied by the network. In cases described by model 12.6b, in fact, an excess input of nutrients is predicted not to change the abundance of phytoplankton (first row, second column). Nutrients could be added to this system for a long time without any apparent change in the abundance of algae, until eutrophication would suddenly appear.[6] On the other hand, if an increase in algae abundance were to be detected, it would immediately be associated with an increased input of nutrients and a management action to keep this variable in check would be called for. The model, on the contrary, tells us that an increase in phytoplankton (F, in the graph) would occur only for parameter changes entering the system through C and Z2 (table of Figure 12.6b, second column, fourth and last row). With no clue about the complex feedbacks that govern the system interventions, based on common sense, straight-line causation would be at the least useless, if not damaging, with economic loss.

The Nile perch (*Lates niloticus*) was introduced for fishing purposes in Lake Victoria during the 1950s and 1960s (Balirwa *et al.*, 1953). A dramatic increase in the stock size of Nile perch roughly coincided with a drastic decline in the population of many indigenous species which ended in a huge biodiversity loss.

---

[6] It has been observed (Jeppesen *et al.*, 1991) that system continuously added with inorganic nutrients may absorb this input without any apparent change. This may continue until a certain threshold is reached when any further addition of nutrient would shift the trophic state to irreversible eutrophy.

Native species, planktivorous in habits, suffered from heavy predation by the Nile perch when already under stress through overfishing. The models in Figure 12.6, in their simplicity, can grasp some of the events that occurred in Lake Victoria.[7] Model 12.6a can describe the core structure of the pelagic community. The introduction of Nile perch can be thought of as producing a negative parameter change on P, whose mortality is increasing. This would reduce the abundance of P itself (table of Figure 12.6a, last row, last column, sign inverted because the input is negative). This new input might have conspired with the already high mortality due to overfishing to make bad conditions for planktivorous fish worse, leading to the extinction of several native populations.

More recently, a resurgence of indigenous species was observed. Overfishing on Nile perch, which in the meantime had become the most preferred species, has been invoked as the main cause of this resurgence. This new preferred stock might have released P from overfishing. These two events may be read as a positive input to P and a negative input to C in model 12.6b, which is now the reference model for the system as it includes the Nile perch (C) as a well-established population. The former input is predicted not to change the abundance of P, with a concomitant increase expected on C (last but one row of the table, last two columns). The negative input on C (overfishing on Nile perch) is expected to reduce C and to increase P (last row, last two columns, inverted signs). Again, the model suggests a chain of events more complex than anticipated: the overall diminution of the Nile perch must be the result of two contrasting effects, such as overfishing and increased availability of prey (planktivorous fish are in fact predicted to increase), of which the former must prevail. However, overfishing on the Nile perch would be the main cause for the concomitant resurgence of indigenous species. The release from fishing pressure, in fact, as a positive input to P is buffered by the network structure and as such no change is expected for P.

## 12.4  Concluding remarks

The fundamental merit of networks is that they educate intuition to cope with complexity. The dominator tree is a vivid example of this. Its algorithm allows unfolding food webs into simpler topologies that are however, not obtained through simplification of the original structure. Rather, they make visible other types of relationships – namely dominance relationships – that are strictly dependent on how the overall energy flow is shaped by the juxtaposition of myriad predator–prey

---

[7] We recognize here that the networks of Figure 12.6 are not detailed enough to embrace the complex dynamics of fish species in Lake Victoria, which is the outcome of the interplay of fine ecological, demographical, and socio-economic mechanisms with their sophisticated feedbacks. Once again we use the models to offer a tool for understanding.

interactions, and which remain hidden in the intricacy of the energy flow map. So dominance relationships are "emergent properties" of food webs that allow one to individuate species that are key players in the flow of energy and whose presence is essential for the survival of other species. With respect to energy flow, thus, dominator trees render keystones species visible, making the arcane obvious.

In a dynamical perspective keystone are often sink species, that is single variables that can be isolable subsystems with zero feedback. For example, a specialized predator that has only one host, no predators, and is not self-damped acts as a sink for its host, like species C in model 12.6b. In these cases, the level of the host population (and that of other populations as well) is buffered against all parameter changes arising elsewhere in the system and responds only to parameter changes entering the system via the predator; also parameter changes entering the system via the host are completely absorbed by the predator.

Beside the role played by the species, food web network analysis distinguishes between interactions: there are in fact functional and redundant links, of which only the former are important to preserve network robustness; the latter can be lost without appreciable consequences (Allesina *et al.*, 2009). In practice, when some functional link is lost the structure becomes more vulnerable to future extinction events with higher probability of cascading extinctions. The importance of links is even more evident in qualitative network analysis. When an ecosystem is perturbed, the impact percolates through many pathways: it may be buffered along some pathways, amplified along others, even inverted on some pathways, giving the opposite result from what is expected. This latter phenomenon is counterintuitive and, as such, it is more interesting to study. It occurs any time a subsystem with positive feedbacks becomes the complement to a certain pathway. These systems with positive feedback may give quite anomalous responses and, therefore, should be searched for in order not to be caught by surprise in management practice concerning biodiversity. If a predator that feeds upon two prey species in competition with one another is subjected to a positive input, for example due to the creation of artificial refugees to increasing its fecundity, the overall effect may be a reduction in its population abundance. In fact the subsystem formed by the two competing prey species has a positive feedback that may invert the effect of the intervention.[8] Overall, networks may help unveiling cause and effect relationships that are confounded by the complexity of ecological systems.

---

[8] This does not necessarily make the whole system unstable as the predator–prey links add further negative feedback to compensate for this positive feedback which, however, remains dominant in the complement for an input on the predator (Bodini, 1990; Bodini *et al.*, 1994).

# 13

# Supply security in the European natural gas pipeline network

MARCO SCOTTI AND BALÁZS VEDRES

Energy security of natural gas supplies in Europe is becoming a key concern. As demand increases, infrastructure development focuses on extending the capacity of the pipeline system. While conventional approaches focus mainly on source dependence, we argue for a network perspective to also consider risks associated with transit countries, by borrowing methods from ecological food web analysis. We develop methods to estimate the exposure and dominance of each country, by using network datasets from the present pipeline system, and future scenarios of 2020 and 2030. We have found that future scenarios will not increase the robustness of the system. Pipeline development to 2030 will shift the relative weight of energy security concerns away from source to transit countries. The dominance of politically unstable countries will increase. The exposure will be slightly redistributed by improving the security of already secure countries, and increasing the exposure of those countries that are already in a vulnerable position.

## 13.1 Introduction

During the first days of 2009 a dispute between Russia and Ukraine led to a closure of major gas pipelines, and the worst dropout of the natural gas supply in Europe so far (Pirani *et al.*, 2009). Supply to 18 countries was disrupted, and some areas with limited reserves and a lack of alternative supply channels were left without heating amidst a bone-chilling cold snap. Initial cuts affected the supplies to Ukrainian consumption (January 1), while deliveries to Europe were reduced drastically on January 6 (e.g. Italy experienced losses of 25% towards its needs and decided to increase imports from Libya, Norway, and The Netherlands; Hungarian consumption was cut off by 40%). The impact was most severe in Central Europe. Thousands of residents in Bulgaria had only an overloaded power grid to rely on

Networks in Social Policy Problems, eds. Balázs Vedres and Marco Scotti. Published by Cambridge University Press © Cambridge University Press 2012.

to keep electric heaters running. Many households in Bosnia were also left without heating, giving a boost to a flu epidemic. Freezing residents in Serbia staged protests against Russia. Croatia and Slovakia declared a state of emergency; much of the industrial production came to a halt. Poland and Hungary also imposed a limit to industrial use of natural gas. Natural gas flows to all European customers did not return to normal levels until January 22.

In this chapter we bring methods and ideas from ecological network analysis – the analysis of food webs – to the problem of energy security in a complex distribution network. The events that unfolded in the crisis of 2009 suggest that supply risks are inherent to the network structure of the natural gas pipeline system.

The supply crisis of 2009 was not an isolated incident. The cascade of energy supply crisis in other similar incidents suggests that the same underlying mechanisms of system fragility are at play. In January 2006, a similar dispute between Ukraine and Russia lasted four days and affected the supplies to many European countries. Italy, for example, reported having lost around 25% of deliveries. Hungary was said to be down by 40% on its Russian supplies (Stern, 2006). In June 2010 Russia cut natural gas supplies via Belarus amidst another dispute. While outside the heating season, this dispute nevertheless affected Lithuanian industrial production as gas imports dropped by 30%.

One ingredient in these repeated supply crises is stress brought about by increasing load on the network system. Consumption of natural gas in Europe has significantly increased over the last two decades. Beyond direct industrial and domestic use, natural gas has also become highly important in electricity generation (Reymond, 2007; Weisser, 2007). The share of gas-fired power generation in all fossil fuels used towards electricity in the 27 current EU member states increased from 11% in 1990 to 34% in 2005, while the amount of electricity produced from coal decreased from 37% in 1974 to 27% in 2004 (IEA, 2007a). According to most major demand forecasts (see for example EU, 2002; IEA, 2007b, p. 492), there is general agreement that this trend will continue in absolute and relative terms.

Another ingredient that contributes towards crises waiting to happen is an ever increasing transportation distance. The fast decline in natural gas reserves near major downstream markets is a serious concern (Lochner and Bothe, 2009). According to a report published in 2007 (WEC, 2007), considering the static range of natural gas (i.e. the time to depletion of proven reserves given current production), conventional domestic reserves may only last for 12 years in North America, and 16 years in Europe. As a contrast, the global expected static range of natural gas is 56 years (WEC, 2007, p. 147). If we look beyond the USA, Europe, and Japan, we see emerging demand centers without significant domestic natural gas reserves (most importantly China and India). These processes make questions of supply security especially acute.

The European pipeline system is intricate in comparison with other energy distribution systems. On a global scale, Europe, USA, and Japan represent the main importers of natural gas. The supply networks to the USA and Japan are far less interesting cases of geopolitical uncertainty than the pipeline network supplying natural gas to Europe.[1] Europe is located proximate to large reserves, and imports natural gas predominantly by pipeline. In 2008 more than 75% of the natural gas consumed by the 27 EU countries was imported. Pipeline imports are "*transported across several sovereign territories*" and "*Each border crossed adds an additional layer of security risk with the potential for conflict within these transit countries, and between the latter and the supplying country*" (Stern, 2002). European countries are strongly dependent on a few exporters for a major part of their gas imports. Eurostat (2008) reports that more than 80% of the EU-27 gas consumption, in 2006, was imported from only three countries: Russia (41.8%), Norway (23.3%), and Algeria (17.5%). In 2005 (IEA, 2007b), Russian natural gas comprised 100% of the imports to many countries (e.g. Finland, Romania, Bulgaria, Slovakia), with a high relative importance also for other EU-27 members (e.g. Hungary 91%, Poland 79%, Austria 78%, Germany 44%).

International trade by ships, for LNG, provides only 15% of natural gas imports to the EU-27 countries (BP, 2008). Pipelines are, and will most likely remain, the main vehicle for natural gas transfer. The peculiar rigidity in the topology of pipelines (i.e. the structure of the distribution system) connecting suppliers (i.e. Russia, Algeria, Libya, Iran, and Azerbaijan) to importing countries deserves attention, in order to investigate its mechanisms of functioning and consequences on supply security.

## 13.2 From food webs to energy supply systems – our network approach to energy security

In this chapter we consider energy security from a network perspective. We study the European pipeline network, a system of pipelines leading to countries in Europe via surrounding countries involved in natural gas distribution (such as Georgia, Syria) and originating from countries in Central Asia, North Africa, or the Middle East. We represent this system in terms of nodes (countries) and links between pairs of nodes (i.e. links summarize the total capacity of pipelines responsible for transferring natural gas across the border of two countries). Our aim is to identify

[1] In 2009 the volume of natural gas imported by USA via pipelines was more than 87.83% of total imports (3,713 bcf supplied by Canada = 87.07% and Mexico = 0.76%), and only 12.17% (452 bcf) as LNG (traded from Egypt = 4.32%, Nigeria = 0.36%, Norway = 0.79%, Qatar = 0.34%, Trinidad & Tobago = 6.36%). These data were extracted from the EIA website: http://tonto.eia.doe.gov/dnav/ng/ng_move_impc_s1_a.htm. Japan is the top LNG importer (with around 3,135 bcf, in 2006). Main sources of the Japanese LNG imports (in 2006) were: Indonesia (22%), Australia (20%), Malaysia (19%), Qatar (12%), Brunei (10%), United Arab Emirates (8%), and Oman (5%). These data were extracted from the EIA website: www.eia.doe.gov/emeu/cabs/Japan/NaturalGas.html.

the role of network topology in security of supply, an approach that is not yet part of the energy policy toolkit (Kruyt *et al.*, 2009).

Network analysis is adopted widely for studying complex systems such as the Internet, motorways, telephone connections, social communities, sexual contacts, or ecosystems (Girvan and Newman, 2002; Liljeros *et al.*, 2001; Watts and Strogatz, 1998). In this chapter we borrow insights from ecological network analysis, as there are many analogies between food webs and the gas pipeline network. Food webs are networks in which nodes correspond to species and links between them stand for energy flows exiting prey items and entering predators (Ulanowicz, 1986). They share many features with the gas transfer network. Firstly, one can trace directed energy flows from sources (i.e. plants converting sunlight into chemical energy) to consumers (e.g. herbivores feeding on plants). Secondly, beyond just classifying node activity into the main roles of prey/source or predator/consumer, the trophic position of a species indicates its relative distance from the primary source of energy (i.e. sunlight), in a longer chain of predation. Thirdly, all nodes (species) in ecological systems show biological respiration (energy dissipation that is not passed on to other species) that is analogous to the domestic natural gas consumption of countries. Fourthly, biological primary producers (plants) rely exclusively on sunlight, without any incoming flows from other system compartments. These are equivalent to natural gas exporters (e.g. Norway, Algeria, and Libya). Finally, the intensity of energy transfers is quantified, in weighted food webs, as energy or matter flows per unit of time and space (e.g. grams of carbon/[m$^2 \cdot$ year]; kcal of energy/[cm$^2 \cdot$ day]). This same approach can be applied on maximal technical capacities of pipelines (measured in normal cubic meters per hour, or Nm$^3$/h).

Our objective is to identify the consequences of source, transit, and structural dependence on natural gas delivery, measuring how indirect effects of a crisis may spread along the pipeline network. We compare risks of countries interrupting gas transfer to the extinction mechanisms in ecosystems. In food web theory there are algorithms to identify bottlenecks (i.e. wasp-waist architecture with one species channeling large amounts of energy; Jordán *et al.*, 2005), and methods for forecasting secondary extinctions (i.e. the possibility that species loss may lead to cascades of further extinctions; Allesina and Bodini, 2004; Dunne *et al.*, 2002b). In this chapter we assess the vulnerability of nodes (exposure) in terms of intensity and number of times by which their natural gas import may be affected. We also estimate secondary losses caused by country exclusion (i.e. a given country switching off natural gas exports), identifying which nodes are likely to cause the greatest impact if removed (dominators).

We aim to answer the following questions: (1) How robust or vulnerable is the configuration of the European gas pipeline network? (2) Will future developments currently planned decrease vulnerabilities in natural gas delivery? (3) How is the political stability of countries related to their pipeline network position and vulnerability of the system?

In the subsequent parts of the chapter we explain the procedures adopted for constructing pipeline networks which represent the architecture of the present scenario (2008), plans that are likely to be realized by 2020, and hypothetical projects that could be operational by 2030. Then, we introduce whole-system indices used for understanding their global functioning. Next, inspired by the concept of trophic position (an ecological index measuring the average distance of a node from the sources of energy), we propose a method for calculating the average number of steps of all the pathways that originate from natural gas sources and reach a given country. After that, we describe simulation techniques to infer exposure (average loss of domestic consumption suffered by closures in the pipeline upstream) and domination (average loss in domestic consumption induced at countries downstream by the closure of pipelines) for each country in the dataset. Finally, we discuss our results in connection with the robustness of future developments in the European pipeline network and its surroundings. We have found that, besides the capability of delivering a greater amount of natural gas, pipelines in the future will also result in few main pathways, increasing system rigidity, and dependence on politically unstable countries. However, we have also observed how more vulnerable countries display lower per capita domestic consumption and will adopt strategies, such as strengthening gas storage facilities, to prevent their exposure.

## 13.3 Data

We analyze the gas pipeline network as composing 55 nodes representing European countries together with Asian and North-African countries on which they rely for consumption (in subsequent sections we use the shorthand "European pipeline network").[2] We describe this system using directed data, depicting graphs of three scenarios. Firstly, we illustrate the structure of pipeline connections in 2008. Then, we define the pipeline architecture as proposed up to 2020. Finally, we represent the network corresponding to long-term plans up to 2030. Countries (nodes) are connected by arrowhead links from exporters towards importers. Link intensities

---

[2] The 55 countries included in the pipeline network and their abbreviations: Albania = AL, Algeria = DZ, Armenia = AM, Austria = AT, Azerbaijan = AZ, Belarus = BY, Belgium = BE, Bosnia and Herzegovina = BA, Bulgaria = BG, Croatia = HR, Czech Republic = CZ, Denmark = DK, Egypt = EG, Estonia = EE, Finland = FI, France = FR, Georgia = GE, Germany = DE, Greece = GR, Hungary = HU, Iran = IR, Ireland = IE, Israel = IL, Italy = IT, Jordan = JO, Kazakhstan = KZ, Latvia = LV, Lebanon = LB, Libya = LY, Lithuania = LT, Luxembourg = LU, Macedonia = MK, Moldova = MD, Morocco = MA, The Netherlands = NL, Norway = NO, Poland = PL, Portugal = PT, Romania = RO, Russia-Kaliningrad = RK, Russian Federation = RU, Serbia = RS, Slovakia (Slovak Republic) = SK, Slovenia = SI, Spain = ES, Sweden = SE, Switzerland = CH, Syria = SY, Tunisia = TN, Turkey = TR, Turkmenistan = TM, UK-Northern Ireland = NI, Ukraine = UA, United Kingdom = UK, Uzbekistan = UZ.

(weights) are measured as cumulated maximal technical capacity ($Nm^3/h$) between two countries.

We record each of the three datasets as a matrix with 55 rows and columns. Element $t_{ij}$ in this matrix corresponds to the cumulated maximal technical capacity if there is, at least, a pipeline from row country $i$ to column country $j$, and 0 if there were none. We have built this network data matrix using a grid map of technically available capacities at cross-border points on the primary market (GIE, 2008b), for EU countries, and other standardized data from specialized websites (Alexander's Gas & Oil Connections,[3] Arab Fund for Economic and Social Development,[4] British Petroleum,[5] Eni,[6] Gazprom,[7] Gulf Oil & Gas,[8] IHS[9]) and literature (EEGA, 2008; EIA, 2008; IEA, 1998; Olcott, 2004; Shirkani, 2008; Yenikeyeff, 2008) to include pipeline details on surrounding countries where main gas reserves are located.

Two 55-element vectors summarize production and current domestic consumption for each node, as reported by *The World Factbook* (CIA, 2008). When volumes of LNG imports are available (BP, 2008), they are summed with the domestic productions. Maximal capacities of infrastructures for LNG imports are extracted from the GIE database (GIE, 2008b, 2010a,b). We consider data on current and future storage facilities (i.e. technical storage capacity, peak withdrawal rate, and peak injection capacity) of European countries, as listed in the GIE website (GIE, 2008a, 2010c,d),[10] and proven gas reserves (CIA, 2008). We also include the following governance indicators developed by the World Bank: rule of law, political stability, and government effectiveness (Kaufmann *et al.*, 2009).

Beyond the network configuration of 2008, we also analyze future scenarios by including new pipelines that are likely to be constructed within 2020

---

[3] www.gasandoil.com

[4] www.arabfund.org

[5] www.bp.com

[6] www.eni.it

[7] www.gazprom.ru

[8] www.gulfoilandgas.com

[9] http://energy.ihs.com

[10] We use information stored into the Excel file describing the GSE Storage Investment Database updated in January 2010 (2010.03.15_Storage Projects Database_Update_Jan2010.xls). In this file, a collection of GSE and non-GSE public information on storage projects around Europe (organized by country and company) is presented. Projects are classified as: (1) LIVE in the case where the facility is operational; (2) planned in the case where the project is at an early evaluation stage; (3) committed in the case where the project is under evaluation by the company with detailed studies and possibly undergoing planning and permitting stages; (4) under construction in the case where the project has already started (procurement and physical operations). Further details are obtained from the GSE_STOR_MapData_August2010_final-1.xls file. This latter summarizes the map of 31 Storage System Operators with around 110 storage sites in 16 countries in Europe, representing approximately 85% of EU technical storage capacity. Using these data sources we are able to define present and future values for: (1) technical storage capacity (million $m^3$) that can be offered to storage users, excluding the capacity for the operational needs of the storage system operator; (2) peak withdrawal capacity (million $m^3$ per day); (3) peak injection capacity (million $m^3$ per day).

and 2030 (Afgan *et al.*, 2007; Mavrakis *et al.*, 2006, pp. 1675–1676; Reymond, 2007; Shirkani, 2008; Yenikeyeff, 2008 – Alexander's Gas & Oil Connections, Arab Fund for Economic and Social Development, British Petroleum, Edison,[11] Eni, Galsi,[12] Gazprom, Hydrocarbons-technology,[13] IHS, Medgaz,[14] Nabucco Gas Pipeline Project,[15] Nord Stream,[16] Trans Adriatic Pipeline,[17] White Stream[18]). Thus, our dataset describes the gas pipeline network at three points in time: the present (2008), the first step of construction (up to 2020), and a second step of construction (up to 2030). In these three networks we preserve the number of nodes ($n = 55$), including new links and adjusting the maximal technical capacity (weight of links) in the case of planned new pipelines and expansion of existing connections. These networks are recorded and visualized by UCINET (Borgatti *et al.*, 2002), a multi-purpose social network analysis software package.

## 13.4 Methods

The first step of our analysis is to compare general indicators about robustness, complexity, and redundancy of the 2008 configuration to the architecture of the plans to be realized by 2020 (step one), and hypothetical projects that could be operational by 2030 (step two). Will these two steps in the development of the European pipeline network bring about a more robust network structure? To answer this question we use indices describing the whole of the network structure, for the present, step one, and step two of future plans. Then, for each country, we aim at investigating vulnerability and domination roles in gas delivery within the present pipeline network and its future configurations. To this end we simulate the effects of country removal on natural gas transfer. Finally, we measure the number of steps separating countries from gas reservoirs or LNG imports. Our goals are to clarify whether planned developments of the European pipeline network will affect energy security, and to identify possible bottlenecks.

### 13.4.1 Whole-system indices

We describe size (total amount of natural gas that can be conveyed using maximal capacities of all the pipelines) and structure (topological constraints to natural gas

[11] www.edison.it
[12] www.galsi.it
[13] www.hydrocarbons-technology.com
[14] www.medgaz.com
[15] www.nabucco-pipeline.com
[16] www.nord-stream.com
[17] www.trans-adriatic-pipeline.com
[18] www.anoproevropu.cz/download/energyForum/vashakmadze.ppt

circulation) of the pipeline network, calculating whole-system indices derived from information theory (Ulanowicz, 2004).

Firstly, we characterize the overall activity of the system by summing the maximal technical capacities of all the pipelines in the network. We call this indicator total system throughput, $TST$.

To estimate system organization we compute the average mutual information ($AMI$ – Rutledge *et al.*, 1976), an amended version of the Shannon's index ($H$ – already applied in other studies as an indicator of energy security – see Jansen *et al.*, 2004; Kruyt *et al.*, 2009). The $AMI$ considers the aggregated constraints exerted upon an arbitrary amount of natural gas when passing from any one country to the next in the pipeline network. The $AMI$ is an index used to define system development as the level of pipeline articulation. The catalyst for its formulation is the Shannon's measure of entropy; the $AMI$ is inversely proportional to the randomness of pipelines, scoring higher when connections are few and/or total maximal capacity distributed is unevenly among them. Both the number of links (pipelines) and their magnitude (maximal technical capacity) determine this index. We measure the $AMI$ to describe the level of pathway multiplicity at disposal of natural gas delivery. The higher the number of alternative routes, the lower the transfer efficiency. In contrast, risks of supply cutoff are strongly attenuated when a certain pathway redundancy is preserved. The performance of a pipeline network may be defined as a tradeoff between transfer efficiency (intended as a few main routes for natural gas circulation) and articulation (responsible for buffering the consequences of perturbations). We adopt the $AMI$ as an aggregated index of energy security, to infer whether the future developments of the European pipeline network will increase its robustness. To this purpose, we also estimate how pipeline connections displayed by each country contribute to its formulation (Scotti *et al.*, 2009). We test differences between $AMI$ values shown by countries, in the three scenarios, using a paired Student's t-test with a two-tailed probability distribution.

To correctly compare the $AMI$ of different networks they should each preserve the same total system throughput ($TST$), which is not realized in our case. As a correction, we use an additional global system index: ascendency ($ASC$), the product of $TST$ and $AMI$. This index gauges system size and distribution of pipeline maximal capacities. It can be interpreted as the tightness imposed on the whole amount of natural gas circulating in the system when it is channeled through the pipeline linkages. Ascendency implies that efficient natural gas delivery is based on rigid network architectures, with a few main trunk lines responsible for the majority of the gas supply. This network efficiency measurement does not focus on the demand for pathway multiplicity, which should be used to capture the robustness of users (countries) to failures in network components (e.g. transmission pipeline breakdowns due to natural disasters, structural accidents, or terrorist attacks). Higher values of ascendency are the consequence of large $TST$ efficiently processed by few and well-structured routes. In this perspective, poorly developed systems possess lower ascendency as an effect of pathway redundancy. This inefficiency, however, may represent an advantage in the case of stress and perturbations (e.g. collapse of pipeline connections caused by natural disasters or disputes between two neighbor countries), providing the system with a degree of adaptability to withstand the consequences of threats and crises (e.g. capturing how countries may dynamically readjust after network disruption, aiming at keeping their consumption patterns constant).

Ascendency is normalized by its upper limit: development capacity ($DC$). This last index describes the entire potential for development of a system and it depends on the observed variety of pipelines, amount of maximal capacity that is available for gas circulation, and the number of nodes. Since $DC$ corresponds to the maximum value that $ASC$ can attain, from their comparison we can make inferences about the developmental status of a network. The highest $ASC/DC$ ratios correspond with the lowest levels of pathway redundancy and highest transfer efficiency.

### 13.4.2 Number of steps from the natural gas sources (NS)

Our baseline expectation is that exposure to supply disturbance is a function of the distance from the natural gas source. We will use this distance as a contrast to measures of exposure that also take network structure into account. We adopt the ecological concept of trophic position ($TP$): the average distance of a node from the sources of energy (Higashi *et al.*, 1989). In food webs, primary producers (plants) receive sunlight energy that they convert into chemical energy. Thus the trophic position of plants is set to 1 ($TP = 1$). As a consequence, herbivores are at $TP = 2$, since energy to them travels a path of length 2 (*sunshine* → *plants* → *herbivores*). Primary predators feeding on herbivores are at the third trophic level, and so forth. In the presence of multiple pathways entering a node (representing a generalist trophic behavior), effective trophic position is calculated partitioning the energy flows into fractions belonging to different trophic levels. In the case of natural gas pipelines, we apply the same algorithm to the pipeline network, defining the number of steps from natural gas sources ($NS$) as the average number of borders crossed by all routes that originate from gas reserves and reach a given country. Countries relying exclusively on gas reserves or LNG imports (considered as input to the system), without incoming pipelines from other nodes, have $NS = 1$ (e.g. Norway, Algeria, Turkmenistan). The $NS$ of countries with incoming pipelines is estimated as the sum of the fractions of integer steps performed by natural gas to reach them. Consider, for example, that the present potential maximal capacity of natural gas to Uzbekistan comes from reserves (7.440 $Nm^3/h$) and Turkmenistan

(5.020 Nm$^3$/h). Uzbekistan has a total input of 12.460 Nm$^3$/h, receiving 59.7% of its income from a one-step pathway (underground $\rightarrow$ Uzbekistan):

$$\frac{7.440 \text{ Nm}^3/\text{h}}{12.460 \text{ Nm}^3/\text{h}} = 0.597, \tag{13.1}$$

and the remaining 40.3% comes from a two-step route (underground $\rightarrow$ Turkmenistan $\rightarrow$ Uzbekistan):

$$\frac{5.020 \text{ Nm}^3/\text{h}}{12.460 \text{ Nm}^3/\text{h}} = 0.403 \tag{13.2}$$

then, the $NS$ of Uzbekistan comes out as follows:

$$1 \cdot 0.597 + 2 \cdot 0.403 = 1.403. \tag{13.3}$$

We adopt a paired Student's t-test with a two-tailed probability distribution for testing the differences between the number of steps separating countries from sources in the three scenarios.

### 13.4.3 Domination and exposure

We run simulations to quantify the direct and indirect effects of the removal of network components (i.e. the case of a country stopping natural gas delivery). The algorithm operates by interrupting, in turn, outgoing flow from each of the 55 countries. Disruptions in gas delivery may lead to direct consequences: absence of any natural gas input and energetic collapse for countries that rely exclusively on pipelines from the removed country; reduction of import potential for countries with multiple inflows including, at least, a pipeline from the country that switched off the natural gas supply. There are also countries without direct links to the node that is removed, but which nevertheless might be affected via their pathway upstream. Consider, for example, a linear pipeline connecting four countries; if the second node of this hypothetical chain interrupts natural gas transfer, both the third and the fourth country would lose all of their natural gas supply, being exposed to direct and indirect effects, respectively.[19] However, when the structure of the network is characterized by some pipeline redundancy, the impact of node removal may only comprise a slight decrease, or no loss, in the natural gas input of the target countries, rather then their collapse.

---

[19] For the purpose of methodological simplification, we do not consider the length of disruption. The timing of response measures by various European countries during the gas crisis between Russia and Ukraine in January 2009 provides relevant examples in this respect. As observed by Pirani *et al.* (2009), various strategies were adopted (i.e. increased imports via alternative pipelines, intensification of LNG trades, gas storage withdrawal, switching to alternative fuels). All of these strategies had a common objective: minimizing dropouts in domestic consumption. However, all these strategies represent short-term solutions (i.e. lasting from 15–120 days), while the presence of multiple pipelines linking a consumer country to different sources provides longer-term stability.

We set two basic rules to simulate natural gas circulation. Firstly, countries aim to preserve their own domestic consumptions. Secondly, they distribute further natural gas, when available, in proportion to the relative importance of their outgoing pipeline capacities. This procedure is illustrated in Figure 13.1.

In the hypothetical network, natural gas is conveyed following the direction of the arrows. We represent the resulting scenario of natural gas distribution if country B becomes isolated from the system: nodes C, D, and E experience a direct loss, but only the latter shows a drop in its consumption (from 30 Nm$^3$/h to 7.9 Nm$^3$/h). These patterns can be explained by taking into account indirect

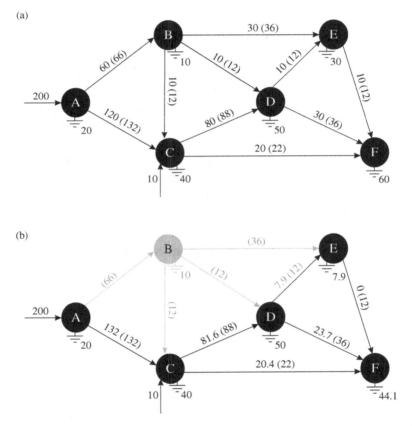

Figure 13.1. Hypothetical network of six nodes: (a) unaffected flow structure; (b) gas transfers and domestic consumptions when the unilateral closure of pipeline valves in B occurs. Maximal technical capacities are indicated as numbers between parentheses next to each arrow. Actual natural gas flows are numbers before maximal capacities. Ground symbols stand for internal consumption. Domestic productions are illustrated with arrows that originate out of no visible node and terminate in receiving nodes (A and C).

effects. In this scenario (Figure 13.1b), imports to C increase by 2 Nm³/h as the removal of B is more than compensated by system resilience through the pathway A → C. Then, natural gas supply to C is rescheduled to 142 Nm³/h (in comparison with the 140 Nm³/h of Figure 13.1a), without affecting its domestic consumption (40 Nm³/h); the remaining amount of natural gas (102 Nm³/h) is distributed to pipelines exiting C, in proportion to their maximal capacity:

$$C \rightarrow D = 102\,\mathrm{Nm^3/h} \cdot \frac{88\,\mathrm{Nm^3/h}}{110\,\mathrm{Nm^3/h}} = 81.6\,\mathrm{Nm^3/h} \qquad (13.4)$$

$$C \rightarrow F = 102\,\mathrm{Nm^3/h} \cdot \frac{22\,\mathrm{Nm^3/h}}{110\,\mathrm{Nm^3/h}} = 20.4\,\mathrm{Nm^3/h}. \qquad (13.5)$$

The same principle is applied in computing the new flows from D to E (7.9 Nm³/h) and F (23.7 Nm³/h), and in deleting the natural gas transfer from E to F. Natural gas transported by the pipeline C → D exceeds domestic consumption of D, but its simulated increase (from 80 Nm³/h to 81.6 Nm³/h) is not enough for buffering (via the route D → E) the supply cutoff to E.

Simulating disruptions in the hypothetical pipeline network (Figure 13.1b) we estimate, for all the nodes, exposure and domination. These are described as impacts on domestic consumption, in accordance with the principles adopted for simulations. We assume that, in the presence of decreased import, a country's internal consumptions tend to be preserved before satisfying gas supply to other countries. Then, a decline in the total input to a country could exclusively result in decreased provision to neighbors, without affecting its own consumption.

Country exposure is measured as the average percentage of domestic consumption losses caused by the removal of upstream countries. Country domination is computed as the ratio between total impacts to all the nodes and the sum of their unaffected domestic consumptions.

For estimating exposure and domination we need data on domestic consumption, impact of country removals, and reachability patterns (the set of countries that lie downstream from a given country in the pipeline). We describe these features using the hypothetical network of Figure 13.1a as a reference. In Figure 13.2a, domestic consumptions are summarized by a column vector (D). The impact matrix [L] (Figure 13.2b) illustrates how row-node removals impact the domestic consumptions of column-nodes. A crisis involving the supply of natural gas passing through node B has consequences on nodes E and F; then nonzero entries in row B correspond to column identifiers of these latter. Finally, the reachability matrix [R] is represented in Figure 13.2c. Reachability is the number of countries (nonzero row values) in all the upstream pathways pointing to the target (column) node (e.g. pipelines to B come only from A and its reachability value is 1, while that of source countries like A is 0). The reachability matrix may also be used to extract nodes in

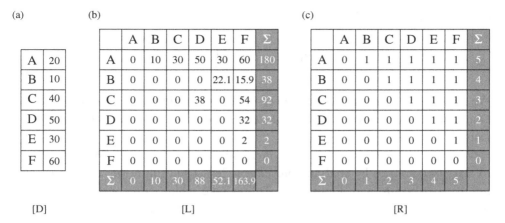

Figure 13.2. Using the network of Figure 13.1a as a reference we illustrate: (a) vector of domestic consumptions (D); (b) impact matrix [L], where the cuts to internal consumptions of column-nodes are caused by gas disruptions in row-nodes ($j$th column sum indicates the total impact on column-node $j$ when, by turn, all the countries arrest gas delivery; $i$th row sum stands for whole reduction of internal consumptions when a gas crisis occurs in the row-node $i$); (c) reachability matrix [R], where upstream (downstream) countries of column (row) nodes corresponds to nonzero row (column) entries. Total upstream and downstream countries are, respectively, column and row sums.

each row-compartment downstream (e.g. node $B$ has a row sum equal to 4, being connected with direct or indirect pathways to all the other system-compartments, except nodes A and B).

Exposure of the column-node $j$ ($\mathrm{Exp}_j$) is computed as the sum of the $l_{ij}$ elements that are listed in the $j$th column of the impact matrix [L], divided by the product of its unaffected domestic consumptions ($d_j$) and total upstream countries ($j$th column-sum of $r_{ij}$ elements extracted from the reachability matrix):

$$\mathrm{Exp}_j = \frac{\sum_{i:i\neq j}^{n} l_{ij}}{d_j \cdot \sum_{i:i\neq j}^{n} r_{ij}}. \tag{13.6}$$

This indicator measures the losses experienced by domestic consumptions of each country when removing, in turn, all the other countries from the network. Effects of the gas crisis occurring in a country may spread beyond neighbor nodes (e.g. indirect consequences on domestic consumptions of node E when the country B arrests natural gas supply – see Figure 13.1).

The extent to which node $i$ behaves as a dominator ($\mathrm{Dom}_i$) is estimated by its downstream effects (i.e. impacts on internal consumptions of the other nodes when $i$ is removed). It is computed by the ratio between row-sum of the $l_{ij}$ elements extracted from the [L] matrix and the sum of all the unaffected domestic consumptions ($d_j$), as listed in (D):

$$\text{Dom}_i = \frac{\sum_{j:j\neq i}^{n} l_{ij}}{\sum_{j:j\neq i}^{n} d_j}. \tag{13.7}$$

We apply the concept of domination to infer impacts due to gas transfer collapse. Domination represents the distribution of losses at all the other nodes in the network caused by the removal of a country.

We use cluster analysis based on increase in sum of squares (Ward Jr., 1963) to classify countries by studying present (in 2008) and future (in 2020 and 2030) trends of exposure and domination. To evaluate which components of the supply security are involved in country exposure and domination, we analyze values extracted from simulations (which take into account constraints of network topology) in comparison with number of steps from sources, domestic production, and proven gas reserves. Next, we correlate exposure, domination, and number of steps from sources with political indicators[20] (rule of law, political stability and government effectiveness), and technological indicators (present – 2008 – and future storage facilities). Then, we investigate which are the relationships connecting the size of gas reserves to indices of political stability. Finally, we test whether there are patterns linking exposure and number of steps from sources to normalized data (by population, GDP, and country area) of domestic consumption. To perform correlation analysis we apply the Pearson's product moment correlation coefficient ($\rho$) for association between paired samples. To carry out simulations and statistical analyses we use the software R (R Development Core Team, 2010).

## 13.5 Results

### 13.5.1 Whole-system performance

European demand for natural gas is expected to increase from 484 billion cubic meters per year (bcm/year) in 2009 to 596 bcm/year in 2030 (Eurogas base case – Marinova, 2010). Determining whether and how the pipeline network is changing to meet these needs is of critical importance for the security of gas supply. The immediate aim of pipeline infrastructure extension projects is to secure the increase in capacities needed to serve this growth in demand. Analyzing the changes projected

---

[20] Using the definition provided by Kaufmann *et al.* (2009, p. 6), rule of law is defined as : "the extent to which agents have confidence in and abide by the rules of society, and in particular the quality of contract enforcement, property rights, the police, and the courts, as well as the likelihood of crime and violence." Political stability is defined as: "capturing perceptions of the likelihood that the government will be destabilized or overthrown by unconstitutional or violent means, including politically-motivated violence and terrorism." Government effectiveness is defined as "the quality of public services, the quality of the civil service and the degree of its independence from political pressures, the quality of policy formulation and implementation, and the credibility of the government's commitment to such policies."

for 2020 and estimated for 2030 we see that total system throughput ($TST$, the sum of all maximal technical capacities occurring in the network) will increase markedly. These future projects will boost the total capacity of the system between 2008 and 2020 by 29.6% (58.82 million $Nm^3/h$). Further development from 2020 to 2030 will result in 25.1% increase (64.53 million $Nm^3/h$). Taking these infrastructure investment steps together, we see a 62.1% increase (123.35 million $Nm^3/h$) between 2008 and 2030. So the pipeline network will most likely serve the increasing demand – but will it become more secure also?

To answer this question we turn to measures developed in ecosystem perspectives: principles adopted in food web theory can be applied to the study of the European gas pipeline network. By analyzing energy exchanges through network architectures, ecologists have been able to unveil how efficiently an ecosystem uses energy and to identify which nodes are likely to cause the greatest impact if removed (Allesina and Bodini, 2004; Dunne *et al.*, 2002b). Achieving energy security in the pipeline network depends on pathway redundancy, to decrease the vulnerability of domestic consumption to perturbations in gas supply. The average mutual information ($AMI$) is an index that captures this structural property (with higher values indicating higher vulnerability).

Considering the changes from 2008 to 2020, and from that intermediate step to 2030, we definitely do not see a decrease in vulnerabilities at the whole-system level. From 2008 to 2020 $AMI$ increases by 9.3%, and from 2020 to 2030, by 2.1%. No significant differences exist between present and future scenarios of gas supply, when the contribution of each country to the whole $AMI$ is considered (all Student's t-tests: $p > 0.05$). So the marked increase in pipeline capacities will not go together with increased topological redundancy.

Further evidence of pipeline network rigidity is provided by the constant ratio between ascendency ($ASC$) and development capacity ($DC$). Ascendency quantifies the level of constraints to the total system capacity, and its upper limit is measured by the development capacity. Higher $ASC/DC$ ratios characterize rigid systems that transfer natural gas via hierarchical pathways. Extremely organized configurations have no redundant connections, a feature that poses serious concerns on energy security. Indeed, redundant trunk lines could serve as degrees of freedom which the system can call upon in the case of disruption (Ulanowicz, 1997), buffering the negative consequences of possible cuts to the gas delivery of upstream countries.

Ascendency of the European pipeline network ranges from 57.5% (2008) to 58.8% (2030) of the development capacity. In spite of the strong increase in total system capacity ($\Delta TST$ between 2008 and 2030 = 62.1%), no significant variations occur in the multiplicity of pathways for gas transfer. Comparison of the ratio between ascendency and development capacity, as measured in the pipeline network, and values displayed by ecological systems can shed light on its meaning. In ecological networks, ascendency of flows ranges between 25% and 50% of the development capacity (Scotti, 2008), while in our study it exceeds 57%. These results corroborate previous outcomes that found human systems to be highly organized (Bodini and Bondavalli, 2002).

Comparing the number of steps separating each country from the natural gas sources ($NS$), in the present and two future steps, we see that more borders will divide gas reserves from consumers in the future. While at present, on average, a country is 2.39 steps from gas reserves, by 2020 this average number of steps will increase to 2.75, and by 2030 it will grow to 2.79. These changes are statistically significant according to paired-sample Student's t-tests: for both tests $p < 0.01$. This means that in the near future a few main pathways (characterized by higher maximal technical capacities) will be responsible for the majority of gas transfers. These main pathways will contribute to vulnerability, as on average there will be more intermediary countries between a consumer and a source country.

In 2008, EU consumers depend on a few sources (i.e. Russia, Norway, and Algeria); some see this as "dangerously too few" (Weisser, 2007, p. 1). Our goal in this part of the analysis is to see whether infrastructure development mitigates this dependence. To estimate vulnerability to supply interruptions we performed simulations in the case of pipeline disruptions by removing each country one by one, and re-calculating the flow. Our goal is to calculate, for each country, the changes in domination and exposure.

### 13.5.2 Changes in domination

Firstly, we examine changes in domination – the influence a country can exert on the natural gas consumption of other European countries. With cluster analysis we classify countries on the basis of their domination at the three points in time that we examine. We consider both 2008 values and expected variations (in 2020 and 2030). Results are summarized by box-plots in Figure 13.3 and Table 13.1. In the four domination clusters, Russia occupies the leading position (D4). The second class of dominators includes both countries increasing (e.g. Turkmenistan, Uzbekistan, and Kazakhstan) and reducing (e.g. Slovakia and Ukraine) their importance, with average impacts on the total European consumptions ranging between 2.5% and 7% (D3). The remaining countries with lower or no effects are grouped in D2 and D1, respectively.

As expected, Russia is the main actor dominating natural gas supply in the European pipeline network, and it will keep its dominant role in the near future of the pipeline system. At present (2008), unilateral closure of pipeline valves by Russia would result in a 23.5% loss for European consumption. Although a

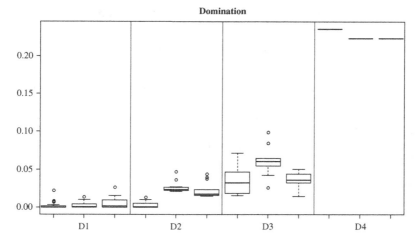

Figure 13.3. Box-plots of countries classified on the basis of current and expected domination. For each group, three box-plots are depicted to describe domination in 2008, 2020, and 2030 (D1 = very low; D2 = low; D3 = intermediate; D4 = high). Values on the y-axis are included between 0 and 1, as they represent relative intensities of domination. Country list in Table 13.1.

slight decrease in this domination is expected with future developments, Russia will maintain its potential to dominate 22.3% of European consumption, both in 2020 and 2030. In the future pipeline configurations, Russia will increase natural gas imports from the Caucasian and Central Asian regions (e.g. Turkmenistan, Uzbekistan, and Kazakhstan), both extending the capacities of existing trunk lines and constructing new pipelines. These countries will extend their domination, both as relative and absolute values, with the potential for impacting more than 6% of European consumption. The new projects that often state an aim to decrease European dependency on Russia (such as the Nabucco, Trans-Adriatic, Greece–Italy, Medgaz, Galsi pipelines) will only achieve this goal marginally.

In Figure 13.4, we illustrate the effects of developments planned for 2020 on the present domination patterns. We depict the expected topology for 2020, with size of nodes proportional to country domination at that step.

As supposed by previous studies (Cetin and Oguz, 2007; Mavrakis *et al.*, 2006), Turkey will challenge the current leading role of Ukraine in natural gas transit to Europe. While Turkey currently does not dominate any consumption of European countries, by 2020 it will dominate 4% of their gas demand. Its relative position made Turkey an ideal candidate to become the main natural gas corridor from Central Asia and the Caspian region to the EU markets, although Cetin and Oguz (2007) observed that this is currently prevented from being the main gate because of the strong dependence on Russia (i.e. natural gas supply is poorly diversified and

Table 13.1. *Dominator countries as classified by clustering analysis (see Figure 13.3).*

| D1 | D2 | D3 | D4 |
|---|---|---|---|
| Albania | Azerbaijan | Algeria | Russia |
| Armenia | Belgium | Germany | |
| Austria | Bosnia and H. | Kazakhstan | |
| Belarus | Bulgaria | Turkmenistan | |
| Croatia | Czech Republic | Tunisia | |
| Denmark | Georgia | Norway | |
| Egypt | Greece | Slovakia | |
| Estonia | Hungary | Ukraine | |
| Finland | Iran | Uzbekistan | |
| France | Macedonia | The Netherlands | |
| Ireland | Poland | | |
| Israel | Romania | | |
| Italy | Serbia | | |
| Jordan | Turkey | | |
| Latvia | | | |
| Lebanon | | | |
| Libya | | | |
| Lithuania | | | |
| Luxembourg | | | |
| Moldova | | | |
| Morocco | | | |
| Northern Ireland | | | |
| Portugal | | | |
| Rep. of Kaliningrad | | | |
| Sweden | | | |
| Slovenia | | | |
| Spain | | | |
| Switzerland | | | |
| Syria | | | |
| United Kingdom | | | |

no significant alternative routes exist for importing gas). Future developments will increase the pipeline redundancy in Turkish import structures (i.e. from Russia, Iran, Azerbaijan via Georgia, and Egypt through an extension of the Arab Gas Pipeline), contributing to a decrease in its vulnerability (Figure 13.4). The future gas pipeline corridor passing through Turkey will benefit from a flexible supply structure, together with higher maximal potential capacity of new infrastructures (e.g. Nabucco Gas Pipeline Project).

Domination patterns under future scenarios also suggest Greece as an emerging bottleneck to natural gas delivery. Its strategic position will be the consequence of new constructions bridging Asian gas producers to Southern Europe demand.

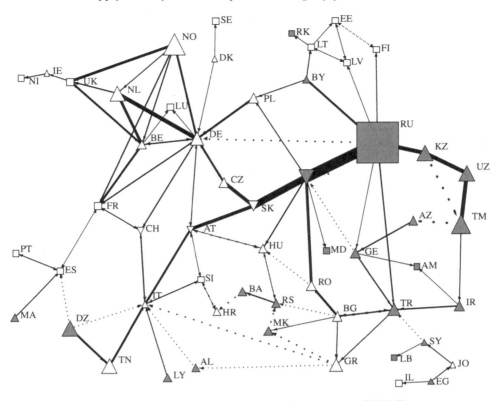

Figure 13.4. Expected pipeline architecture and domination in 2020. The presence of pipeline connections between pairs of nodes, as forecasted for the 2020 scenario, is illustrated (2008 pipes are solid links; new pipes that will be operational in the 2020 network are dotted lines). Natural gas direction is indicated by arrows. Link thickness is proportional to the cumulated amount of natural gas that is expected to be transported by pipelines in 2020 (measured in $Nm^3/h$). The size of the nodes (countries) reflects their domination (the bigger the node, the higher are the losses in the domestic consumption of the other countries, in case the target node interrupts gas delivery). The shape of nodes represents the changes in 2020 domination, in comparison with 2008: increasing and decreasing patterns with percentages higher than 15% are represented by upper and lower triangles, respectively; squared nodes indicate weaker fluctuations. Shading of nodes describes country political stability (grey nodes have a sum of three scores – political stability, government effectiveness, and rule of law – less than 50% of the maximum).

Greece will play two different roles: firstly, representing an alternative by-pass to the Turkish corridor, and conveying Russian natural gas (received through Bulgaria by means of the South Stream pipeline); secondly, distributing to Italy natural gas received from Turkey (Interconnector Turkey–Greece–Italy). Similarly to Turkey, the dominance of Greece is strengthened by very low levels of exposure.

Table 13.2. *Correlation coefficients of log domination versus number of steps from natural gas sources (NS), log size of gas reserves (million $m^3$), and log domestic production (bcm/year).*

|  |  | NS | | Gas reserves | | Domestic production | |
|---|---|---|---|---|---|---|---|
|  |  | $\rho$ | p | $\rho$ | p | $\rho$ | p |
| **Domination** | 2008 | −0.293 | 0.116 | 0.529 | 0.005 | 0.568 | 0.003 |
|  | 2020 | −0.072 | 0.656 | 0.380 | 0.022 | 0.288 | 0.093 |
|  | 2030 | −0.144 | 0.357 | 0.311 | 0.065 | 0.233 | 0.177 |

Another interesting pattern characterizes the domination of transit countries between the Russian and the Caspian Sea region sources (i.e. Turkmenistan, Kazakhstan, and Uzbekistan), and Western European consumers. The majority of Eastern European (e.g. Bulgaria, Romania), Central European (e.g. Hungary, Czech Republic, Poland) and Balkan countries (e.g. Serbia, Croatia, Bosnia and Herzegovina) will increase their domination, while the current main corridor conveying natural gas from Russia through Ukraine and Slovakia will become less important. Significant changes will also take place in the pathway for transporting natural gas towards Turkey and Greece, reflecting the aspiration of the former to become the main Eurasian natural gas pipeline corridor. Also North African countries (e.g. Libya and Algeria) will display higher domination levels.

Is domination a simple function of countries being closer to natural gas reserves? If yes, then there is no complexity in answering the question of what makes countries dominant, and how to mitigate dependency on them. Considering the correlation between our measure of domination and the number of steps from the source, we can reject the hypothesis that domination is a simple matter of being upstream (see Table 13.2). There is no significant correlation between domination and the number of steps from the source, which suggests that domination is more a function of the topology of the network.

Is domination only a function of countries having gas reserves themselves? Again, if the answer is yes, there is no need to consider the network topology of pipelines. We see in Table 13.2 that the correlation between domination and gas reserves is significant and positive at present, although the correlation is considerable smaller than one. However, as we consider plans into the future, the correlation is decreasing, and by 2030 it is not significant at the $p < 0.05$ level. The same trend is found when we analyze domination in comparison with domestic production: the significant correlation observed for 2008 ($\rho = 0.568$, $p = 0.003$) vanishes in future scenarios (see Table 13.2). We also consider this latter hypothesis because, in some

Table 13.3. *Comparison between indicators of political stability, log domination, and log size (in bcm) of gas reserves.*

|  |  | Rule of law | | Political stability | | Government effectiveness | |
|---|---|---|---|---|---|---|---|
|  |  | $\rho$ | p | $\rho$ | p | $\rho$ | p |
| **Domination** | 2008 | −0.022 | 0.910 | 0.115 | 0.546 | 0.000 | 0.999 |
|  | 2020 | −0.346 | 0.027 | −0.228 | 0.152 | −0.215 | 0.177 |
|  | 2030 | −0.324 | 0.034 | −0.298 | 0.052 | −0.277 | 0.072 |
| **Gas reserves** |  | −0.348 | 0.032 | −0.219 | 0.186 | −0.361 | 0.026 |

cases, the existence of reserves and pipelines does not mean that the gas is delivered (e.g. Iran has the second largest reserves in the world but is still a net importer). Our findings indicate that the development of the pipeline system will shift security concerns from source countries to transit countries. In the future, domination will be more based on network position, rather than on gas reserves.

Thus far we have considered domination equally for all countries. For example, closure of all pipelines by Ukraine would mean a 7.2% loss to European consumption, and pipeline closure by Norway would mean a similar 6.3% loss to European consumption. European consumption is less at risk from Norway, than from Ukraine: not all countries are equally likely to translate their dominance into energy security concerns. To incorporate this idea in our analysis we considered the correlation between rule of law, political stability, and government effectiveness (Table 13.3).

In the present (2008) natural gas pipeline system there is no correlation between rule of law and domination. That is, the rule of law in countries with potentially high impact on the domestic natural gas consumption of other downstream countries is not different from the average rule of law of all countries. However, in 2020 and 2030 this correlation will become negative and statistically significant: countries with higher domination in the future will also have a weaker rule of law. Similarly with political stability, there is no correlation with domination in the present system, but by 2030 it is becoming a significant negative correlation ($p = 0.052$). Government effectiveness works in a similar way, though the correlation is slightly below the threshold of 0.05 ($p = 0.072$). This suggests that, in the future development of the system, domination will shift to politically unstable countries with a weak rule of law. We also note that there is a correlation between natural gas reserves in a country and poorer scores on these three dimensions of institutional stability (Table 13.3). This finding is in line with the expectation that abundance of natural resources would go together with poorer developmental outcomes and poorly performing state institutions (Karl, 1997). As argued by Corden and Neary (1982), discovery and exploitation of large gas reserves may have negative impact on the traded sector, making the other products of the country less price competitive on the export market. This mechanism, also known as "Dutch disease problem," may lead to negative growth effects, with the reallocation of production factors apart from resource exploitation.

### 13.5.3  Changes in exposure

We now turn to analyzing the exposure of domestic consumption in each country to interruptions in the pipeline system by other countries. Considering change in exposure between 2008 and 2020, we see that there are 15 countries where exposure will decrease, but there are 16 countries where exposure will increase, as a result of new developments. As we consider infrastructure work to be completed by 2030, we see a little more positive distribution, with 20 countries improving on their present exposure, and only 11 countries becoming more vulnerable. However, the improvements in exposure tend to be very small, and the increases in exposure tend to be much more pronounced.

To describe typical trends across exposure at three points in time, we identify four clusters (Figure 13.5 and Table 13.4). Most of the countries, more than two thirds,

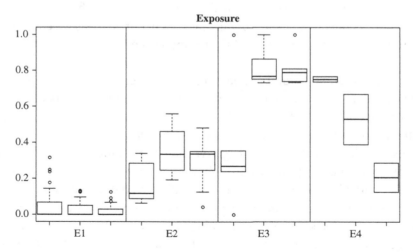

Figure 13.5. Present and future exposures are used to classify countries in four categories. For each group, we depict box-plots of 2008, 2020, and 2030 exposure values (included between 0 and 1). We identify countries preserving low vulnerability (E1), increasing exposure (E2 and E3) and strengthening stability (E4). The full list of countries is provided in Table 13.4.

Table 13.4. *Country exposure as classified by cluster analysis (see Figure 13.5).*

| E1 | E2 | E3 | E4 |
|---|---|---|---|
| Albania | Austria | Ireland | Bosnia and H. |
| Algeria | Belarus | Northern Ireland | Estonia |
| Armenia | Croatia | Slovenia | |
| Azerbaijan | Czech Republic | Sweden | |
| Belgium | Hungary | Switzerland | |
| Bulgaria | Lithuania | | |
| Denmark | Moldova | | |
| Egypt | Slovakia | | |
| Finland | Tunisia | | |
| France | | | |
| Georgia | | | |
| Germany | | | |
| Greece | | | |
| Iran | | | |
| Israel | | | |
| Italy | | | |
| Jordan | | | |
| Kazakhstan | | | |
| Lebanon | | | |
| Libya | | | |
| Luxembourg | | | |
| Morocco | | | |
| The Netherlands | | | |
| Norway | | | |
| Portugal | | | |
| Rep. of Kaliningrad | | | |
| Russia | | | |
| Spain | | | |
| Syria | | | |
| Turkey | | | |
| Turkmenistan | | | |
| Ukraine | | | |
| United Kingdom | | | |
| Uzbekistan | | | |
| Latvia | | | |
| Macedonia | | | |
| Poland | | | |
| Romania | | | |
| Serbia | | | |

Table 13.5. *Correlation coefficients of log exposure versus number of steps from natural gas sources (NS) and log normalized consumptions. Domestic consumptions are normalized by the ratio with number of inhabitants (Population – # of inhabitants), gross domestic product (GDP – $), and land area (area – km².*

| | | NS | | Population | | GDP | | Area | |
|---|---|---|---|---|---|---|---|---|---|
| | | $\rho$ | p | $\rho$ | p | $\rho$ | p | $\rho$ | p |
| **Exposure** | 2008 | 0.193 | 0.289 | −0.395 | 0.031 | −0.202 | 0.284 | −0.253 | 0.178 |
| | 2020 | 0.387 | 0.026 | −0.431 | 0.017 | −0.127 | 0.505 | −0.006 | 0.973 |
| | 2030 | 0.303 | 0.092 | −0.472 | 0.010 | −0.073 | 0.705 | −0.024 | 0.902 |

fall into our first cluster, E1. These countries start with a low level of exposure in the present: on average only about 4% of their domestic consumption is at risk from pipeline closure upstream. Infrastructure development will improve their low exposure, on average by two percentage points. The second cluster, E2, with nine countries starts from an exposure of 18% on average, which increases considerably, by about 12 percentage points by 2030. The third cluster, E3, repeats this pattern: five countries start from an average exposure of 37%, which increases dramatically, to 81% on average. We only found two cases, Estonia, and Bosnia and Herzegovina, our forth cluster, E4, where pipeline development will result in a major decrease in vulnerability, from 75% to about 20%.

Pipeline development will not eliminate country exposures – but will mostly redistribute it. Improvements are expected for those countries that have low exposures, while countries with higher exposures tend to become more exposed. Only development projects in the more distant future seem to bring a significant decrease in the number of exposed countries.

Exposure is not a simple function of the number of steps from natural gas sources: this correlation is not significant in the 2008 and 2030 scenarios (Table 13.5). However, we observe that a significant correlation between exposure and number of steps from natural gas sources is attained by 2020 ($\rho = 0.387$, $p = 0.026$). Countries with higher exposure (Exp $\geq 0.50$) tend to have lower than 0.6 million Nm³/h natural gas consumption per capita, which probably reflects longer-term (at least five years) adjustment of the energy mix to energy security concerns (Table 13.5).

No correlation is observed when comparing indicators on present and future storage facilities (GIE, 2010a,b,d) with current exposure (Table 13.6). This finding corroborates the idea that considerations on energy security are of secondary importance when planning the construction of new infrastructures. The main reason stimulating storage investment is to match inflexible production with fluctuating

Table 13.6. *Correlation coefficients of (present and future) log normalized storage size versus log exposure and number of steps from natural gas resources (N S). Storage facilities are normalized by the ratio between storage size (million $Nm^3$) and domestic consumption (million $Nm^3/year$).*

| | | Storage facilities | | | |
|---|---|---|---|---|---|
| | | present | | future | |
| | | $\rho$ | p | $\rho$ | p |
| **Exposure** | 2008 | 0.373 | 0.232 | 0.291 | 0.335 |
| | 2020 | | | 0.717 | 0.004 |
| | 2030 | | | 0.668 | 0.009 |
| *N S* | 2008 | 0.474 | 0.054 | 0.358 | 0.144 |
| | 2020 | | | 0.465 | 0.052 |
| | 2030 | | | 0.514 | 0.029 |

consumption. Demand for natural gas varies depending on the season and time of day; for technical reasons, it is not possible to balance production and import to fluctuating demand. Production and import remain essentially constant throughout the year, while seasonal and trading-related time factors characterize the internal consumption. Then, it is efficient to build new storage, or extending the capacity of an existing facility, based on the magnitude of the domestic demand. This clearly explains the strong correlation linking present consumption patterns (CIA, 2008) to storage capacity ($\rho = 0.581$, $p = 0.001$), withdrawal rate ($\rho = 0.687$, $p = 0.002$), and injection capacity ($\rho = 0.543$, $p = 0.024$) of facilities that were operational in 2008. These trends are further strengthened when plotting current domestic consumption against technical properties of future storage facilities (i.e. taking into account planned, committed, and under construction projects). Strong correlations are found between current consumptions and expected storage capacities ($\rho = 0.864$, $p < 0.001$), withdrawal ($\rho = 0.683$, $p = 0.001$), and injection ($\rho = 0.613$, $p = 0.005$) rates. The role of storage facilities in compensating variations between supply and demand may also help to explain the correlation between the number of steps separating countries from natural gas sources and storage capacity (Table 13.6). Highest levels of uncertainty are associated with the regularity of natural gas supply when many borders have to be crossed for its delivery.

## 13.6 Conclusions

Adopting a network perspective, we develop an integrated approach to energy security concerns of availability (i.e. elements related to geological existence and location of gas reserve – source dependence) and accessibility (i.e. natural gas transportation which often involves geopolitical implications, leading to transit dependence). Network analysis considers features such as import dependence, source and transit diversity, both at the whole-system level (e.g. $TST$, $AMI$, $ASC/DC$) and at the country level (e.g. $NS$, exposure, domination). For a broader understanding of European gas security, we have also extended our study by including political stability and storage facilities. With this approach we aim to address energy security as a multifaceted concept (Kruyt *et al.*, 2009): to develop metrics useful in policy-making contexts (allowing the setting of targets in a similar way to setting greenhouse gas reduction goals), multiple dimensions of energy security need to be captured.

Our findings are paradoxical in several respects. Contrary to widespread expectations that pipeline development projects will increase gas security, we find that the European pipeline network is evolving towards a more fragile configuration. Moreover, there is evidence for an increasing dependence on politically unstable countries, such as Russia, Ukraine, Turkey, Algeria, and Libya. Considering expected pipeline developments from 2008 to 2020 and 2030, we will see a restricted number of routes transporting higher amounts of natural gas, and crossing several national borders, with the emergence of new bottlenecks, threatening the security of supply (Stern, 2002).[21] Besides Ukraine and Belarus, two main transit constraints in the 2008 pipeline configuration, other countries (e.g. Greece, Turkey, Bulgaria, Romania, and Poland) will emerge as key players for gas delivery in 2020 and 2030. They will occupy strategic positions in future routes for gas transportation (i.e. Nabucco, South Stream, Poland–Czech Republic Interconnector, Trans-Adriatic pipeline). Infrastructure development seems mainly focused on increasing the range of choice with respect to the gas reserves reached. In spite of projects explicitly aimed at avoiding transit problems (consider the pipelines Nord Stream and South Stream, for example), our findings indicate that transit risks will not be less acute. Although Russia will remain the main actor, higher amounts of natural gas will be supplied by other source countries (e.g. Central Asian and North African). The current strong correlation between domination and both gas reserves and domestic production will mitigate, and domination will shift towards transit countries. For example, the Turkey–Greece corridor will challenge the current

---

[21] Future projects especially of interest here are those that will entail the construction and expansion of pipelines transporting amounts of natural gas ranging between 1 million normal cubic meters per hour ($Nm^3/h$) and 3 million $Nm^3/h$. Some of these routes will aim at increasing imports from the North African area to European countries (e.g. the GALSI pipeline that is expected to be operational by 2014, for transporting natural gas from Algeria to Sardinia, and further to Northern Italy). Other projects are planned for connecting Caucasian and Central Asian regions to the European market: South Stream (Russia → Bulgaria → Republic of Serbia → Hungary (→ Slovenia) → Austria; Russia → Bulgaria → Greece → Italy; due by 2015) and Nabucco (Turkey → Bulgaria → Romania → Hungary → Austria; operational by 2015) pipelines.

leading role of the Ukrainian gate for natural gas transfers from Asian resources to the EU market.

Investment for construction of new pipelines or extending the capacities of existing trunk lines will not improve radically the current patterns of exposure for many countries. In particular, with the pipelines that are likely to be operational by 2020 and 2030, just Estonia (Exp-2008 = 0.76; Exp-2030 = 0.13) and Bosnia and Herzegovina (Exp-2008 = 0.74; Exp-2030 = 0.29) are expected to strongly reduce their exposure. Countries falling into the category E1 (Figure 13.5) are not affected by high exposure, either in the present or in the future scenario (avg Exp-2008 = 0.05; avg Exp-2030 = 0.02). However, 14 countries out of 55 (categories E2 and E3, Figure 13.5) will be exposed to higher transit risks as a consequence of future pipeline configurations (2020 and 2030). In the current network (2008), they already display concerns on energy security (avg Exp group E2 = 0.18; avg Exp group E3 = 0.37), but they will be further endangered by the developments due by 2030 (avg Exp group E2 = 0.30; avg Exp group E3 = 0.81).

Storage facilities do not represent a comprehensive solution to the issues of transit and source dependence. However, they can be effective in withstanding the consequences of a short-term cutoff (Weisser, 2007). Consider, as an example, the role played by natural gas storage facilities during the Russian–Ukrainian crisis of January 2009. Some European countries used alternative fuels for satisfying natural gas requirements of industrial activities (e.g. Bulgaria, Croatia, Bosnia and Herzegovina), but in other cases storage facilities were able to cover demand for several weeks (e.g. Austria, Hungary, Poland, Germany).

Our analysis is open to improvement through further investigation. Firstly, to shed light on facility dependence (Luciani, 2004), the European gas transit network could be detailed at the level of particular pipelines (as opposed to summing pipeline capacities for country pairs). One could also increase network resolution by describing the structure within countries, including routes connecting facilities to consumer nodes. Secondly, differences in tariffs could be used to simulate the most likely gas routing within the network (Pelletier and Wortmann, 2009). Thirdly, one could learn a lot by comparing the carbon budgets (for all energetic consumption) with country exposure. This approach could also address the issue of climate change jointly with energy security in the same framework (Kruyt *et al.*, 2009). Finally, the main objective of this chapter concerns security related to nodes (countries), but an alternative perspective that considers system fragility as a function of connection removal deserves attention for further studies. This would allow description of system vulnerability as depending on the failure of specific infrastructures. Allesina *et al.* (2009) proposed a method for discriminating between functional and redundant links in food webs. The authors focused on the role of trophic connections, rather than investigating secondary extinctions triggered by node (species) removal.

They showed that trophic links can be divided into functional and redundant, based on their contribution to robustness. They also found that food webs are characterized by hubs (the most connected species) that are not necessarily the most important species, as they may hold many redundant links. Similarly, security rules for the electric power grid are based on the idea that the system should be able to withstand the loss of any single connection. When a power system satisfies this criterion, it is said to be "N-1 secure," because it could lose any one of its N lines and continue operating. In a system that is "N-2 secure," no consumer would be disconnected even if two connections were suddenly removed (Song and Kezunovic, 2007).

# 14

# Conclusions and outlook

ALBERT-LÁSZLÓ BARABÁSI

## 14.1 Network science: roots and perspectives

Graphite is a soft, easy to break material, whose physical properties are rooted in the weak coupling between the layers its atoms form. In contrast diamond, whose atoms are arranged in a face-centered cubic lattice, is the hardest material we know on Earth. But the same atoms can also form a hollow sphere, called buckyball, or a cylinder that goes by the name of carbon nanotube. These materials have such fascinating chemical properties that their discovery was honored with a chemistry Nobel Prize in 1996. When these atoms form stacked sheets of linked hexagonal, pentagonal, or heptagonal rings, we call them graphene, with their own impressive list of properties and potential applications, from nano-transistors to gas detectors, a reason why this material swept the 2010 physics Nobel Prize.

What is fascinating about this list is that each of these materials, with such diverse appearance and physical and chemical characteristics, is made of the same atom: carbon. Yet, at the end there is only one difference between graphite, graphene, diamond, buckyballs, carbon nanotubes, and many other allotropes of carbon: the shape of the network that links the carbon atoms to each other. Which is an observation that increasingly resonates well beyond physics and chemistry: when it comes to the society, the economy, or even our cells, some tough, others malleable, some alive while others barely hanging on, many of their characteristics are rooted in the structure of the network that waves their building blocks together.

We are surrounded by systems that are hopelessly complicated, from society, a collection of seven billion individuals, to communications systems, integrating billions of devices like computers and cell phones. Our life relies on the ability of thousands of genes to work together in a seamless fashion. Our thoughts, reasoning,

---

Networks in Social Policy Problems, eds. Balázs Vedres and Marco Scotti. Published by Cambridge University Press © Cambridge University Press 2012.

and our ability to comprehend the world that surrounds us are conditioned by the coherent activity of billions of neurons. Our financial health depends on the seamless cooperation of banks, companies, and governments. The details of these systems are overwhelming, so much so that many appear to be random to the casual observer. Upon closer inspection however, they display, endless signatures of order and self-organization. They are collectively called *complex systems*, and given the important role they play in our life, their quantitative understanding, prediction, and eventual control is becoming the scientific challenge of the twenty-first century.

This book is an excellent illustration of the breadth and depth of the challenges that complex systems and networks offer to us. By exploring socio-economic problems from a network perspective, touch on a number of subjects at the forefront of social sciences, from the spread of information and ideas to the impact of networks on entrepreneurship and creativity; from the operation of institutions to macro-level crises. Together with similar recent volumes that focus on biological networks and communication systems as networks, this shows how network thinking fundamentally alters traditional disciplines. This explosion of interest raises an important question that we need to address: why now? Why did network science not emerge two hundred years earlier?

The subjects of network science are by no means new – the metabolic network is four billion years old, dating back to the origins of life. The World Wide Web has been around for two decades. The only young kid on the block is Facebook, not having yet reached its teenager years, but one could argue that the social network it captures goes back several millennia. Scientific interest in these systems is by no means new either: metabolism has been studied intensively for about 150 years, the web has had its share of scrutiny over the past two decades, and despite its tender age, already volumes of research focus on Facebook. Furthermore, many disciplines, from sociology to brain science and graph theory, have been dealing with their notion of networks for many decades, if not centuries. So why do we often call network science the science of the twenty-first century?

The reason is simple: something special happened at the dawn of the century that fueled the emergence of a new research field with its own intellectual challenges. To understand why this happened now, and not two hundred years ago, we need to highlight a few prominent forces.

### Data availability

To understand a complex system, we need a map of the system's wiring diagram. In a social system, this would equate to knowing everyone's list of friends. In the web, this is a map telling us which webpages link to each other. In the past we either lacked the technology to map out these networks, or it was impossible to store and organize the huge amount of data these maps represent. The Internet, abundant computing power, and cheap digital storage have fundamentally changed this situation, leading to the emergence of a series of maps of real systems that have both shaken and enriched our understanding of networks. Interestingly, many of the "canonical" network maps that served as the starting point of discoveries that changed network science over the past decade were not collected for the purpose of studying networks. The lists of chemical reactions that take place in a cell were uncovered over a 150-year period by biologists, trailed by numerous Nobel prizes. Yet, we had to wait for the World Wide Web to make these widely accessible in central databases. The list of actors that play in each movie was available in dusty books and encyclopedias. We needed imdb.com, fueled by the curiosity of movie aficionados, to make these lists accessible in a format that allowed researchers to analyze them. In the hands of network scientists, these databases turned into network maps: the list of metabolic reactions became the metabolic network, helping us to discover the small-world nature of cellular networks; the map behind Hollywood led to the discovery of network clustering. Other maps were collected with the explicit goal of understanding the properties of real networks, like the first publicly available WWW map obtained by Hawoong Jeong, helping the discovery of scale-free networks. Today, recognition of the important role networks play in understanding complex systems fuels a wide array of mapping projects. Examples include major programs aimed at obtaining an accurate map of the Internet, the hundreds of millions of dollars spent by biologists to map out protein and regulatory interactions in various organisms, or the connectome project, with the ambitious goal of producing a neural map of a mammalian brain.

## Universality

It is easy to highlight substantial differences between the metabolic network, whose purpose is to process energy in the cell, and the World Wide Web, whose purpose is information access. The nodes of the metabolism are tiny molecules (metabolites) and links are chemical reactions. The nodes of the web are web documents and the links are the URLs. The forces and processes that shape these networks also differ greatly: metabolic networks are shaped by billions of years of evolution, while the web was built by humans over a mere 20-year period, and we have only one copy of it. Given this diversity in size, nature, scope, history, and evolution, one expects that the networks behind these two systems should differ greatly. Yet, perhaps the most important discovery of network science is that there are fundamental similarities between these and many other networks emerging in various domains of science, nature, and technology. Such similarities, often called organizing principles, like the scale-free property or degree correlations, have helped us to develop an ability to observe and quantify the common and the generic features of apparently dissimilar systems, leading to a series of fundamental laws that govern network evolution.

## *The need to understand complexity*

Modern science is built on the reductionist paradigm, a belief that we can best understand the nature of a system by reducing it to its parts. The basic hypothesis behind the reductionist paradigm is that once we understand the pieces, we can rebuild and understand the whole system. But the reassembly has turned out to be much harder than scientists had anticipated. We have learned that, in complex systems, the components can fit in so many different ways that it would take billions of years for us to try them all. In spite of the challenges these systems pose, we cannot afford to not address their behavior, a view increasingly shared by both scientists and policy-makers, as these systems are at the heart of our economy, our technologies, and our society. Networks are not only essential for this journey, but during the past decade some of the most important advances towards understanding complexity have been provided in the context of network theory, fueling continued interest in this area.

The impact of these three forces is well illustrated in the previous chapters: the papers are predominantly empirical, relying on datasets that were either collected specifically for these projects or piggybacked on existing datasets. The insights they offer about social and economic systems reinforce the value of the data-oriented nature of network science. Most of these papers, implicitly or explicitly, make use of universality, taking the generic patterns that pertain to network spreading, robustness, and fragility, to mention just two, in new directions. They also speak the universal language of networks, the one that has lately allowed sociologists, biologists, physicists, and computer scientists to converse with each other, sharing tools, ideas, and concepts. Finally, they address issues of major societal and economic importance, underlying the third leg of network science: the desperate need to understand these complex systems.

Network science is by no means the answer to all of the questions that complex systems challenge us with. There is so much more to complexity than a network describing who is interacting with whom. Yet one thing is clear: it would be delusional to believe that we can understand complex systems without fully understanding the networks behind them. Networks are such an essential component of this journey, that no theory or approach that aims to account for their behavior can afford to by-pass their understanding.

# References

Abbott, A. 1995. Things of Boundaries. *Social Research*, **62**, 857–882.

Adams, J. D., Black, G. C., Clemmons, J. R., and Stephan, P. E. 2005. Scientific teams and institutional collaborations: evidence from U.S. universities. *Research Policy*, **34**, 259–285.

Afgan, N. H., Carvalho, M. G., Pilavachi, P. A., and Martins, N. 2007. Evaluation of natural gas supply options for south east and central Europe. Part 1: indicator definitions and single indicator analysis. *Energy Conversion and Management*, **48**, 2517–2524.

Agency for Toxic Substances and Disease Registry. 2007. *Case studies in environmental medicine (CSEM) nitrate/nitrite toxicity.*

Aksnes, D. W. 2006. Citation rates and perceptions of scientific contribution. *Journal of the American Society for Information Science and Technology*, **57**, 169–185.

Albanese, D., Merler, S., Jurman, G., Visintainer, R., and Furlanello, C. 2009. *MLPY Machine Learning PY*. http://mloss.org/software/view/66/.

Albert, R. and Barabási, A. -L. 2002. Statistical mechanics of complex networks. *Reviews of Modern Physics*, **74**, 47–97.

Albert, R., Jeong, H., and Barabási, A. -L. 2000. Error and attack tolerance of complex networks. *Nature*, **406**, 378–381.

Albert, R., Albert, I., and Nakarado, G. L. 2004. Structural vulnerability of the North American power grid. *Physical Review E*, **69**, 103–107.

Alcácer, J. and Gittleman, M. 2006. Patent citations as a measure of knowledge flows: the influence of examiner citations. *The Review of Economics and Statistics*, **88**, 774–779.

Alexa, N., Bárdos, R., Szántó, Z., and Tóth, I. J. 2008. Corruption risks in the business sector. Integrity system country study (part two). *Transparency International*.

Ali, S. N., Young, H. C., and Ali, N. M. 1996. Determining the quality of publications and research for tenure or promotion decisions: a preliminary checklist to assist. *Library Review*, **45**, 39–53.

Allard, A., Noël, P. -A., Dube, L., and Pourbohloul, B. 2009. Heterogeneous bond percolation on multitype networks with an application to epidemic dynamics. *Physical Review E*, **79**.

Allesina, S. and Bodini, A. 2004. Who dominates whom in the ecosystem? Energy flow bottlenecks and cascading extinctions. *Journal of Theoretical Biology*, **230**, 351–358.

Allesina, S., Bodini, A., and Bondavalli, C. 2006. Secondary extinctions in ecological networks: bottlenecks unveiled. *Ecological Modelling*, **194**, 150–161.

Allesina, S., Bodini, A., and Pascual, M. 2009. Functional links and robustness in food webs. *Philosophical Transactions of the Royal Society of London – Series B: Biological Sciences*, **364**, 1701–1709.

Amaral, L. A. N. 2005. Novel collaborations within experienced teams lead to best research outcomes. *Annals of Vascular Surgery*, **19**, 753–754.

Amaral, L. A. N. and Ottino, J. 2004. Complex networks: augmenting the framework for the study of complex systems. *European Physical Journal B*, **38**, 147–162.

Amaral, L. A. N. and Uzzi, B. 2007. Complex systems – a new paradigm for the integrative study of management, physical, and technological systems. *Management Science*, **53**, 1033–1035.

Amaral, L. A. N., Scala, A., Barthélémy, M., and Stanley, H. E. 2000. Classes of small-world networks. *Proceedings of the National Academy of Sciences U.S.A.*, **97**, 11149–11152.

Amenta, E., Carruthers, B., and Zylan, Y. 1992. A hero for the aged? The Townsend Movement, the political mediation model, and U.S. old-age policy, 1934–1950. *American Journal of Sociology*, **98**, 308–339.

Andrews, K. T. and Edwards, B. 2005. The organizational structure of local environmentalism. *Mobilization*, **10**, 213–234.

Andvig, J. C. and Fjeldstad, O. -H. 2000. *Corruption. A review of contemporary research. NUPI Report No. 268*. Tech. rept. Oslo: Norwegian Institute of International Affairs. Also published as CMI Report R 2001:7.

Ansell, C. 2003. Community embeddedness and collaborative governance in the San Francisco Bay Area Environmental Movement. Pages 123–144 of: *Social Movements and Networks*. Oxford University Press.

Armstrong, E. A. 2005. From struggle to settlement: the crystallization of a field of lesbian/gay organizations in San Francisco, 1969-1973. Pages 161–187 of: *Social Movements and Organizations*. Cambridge University Press.

Aronowitz, S. and Bratsis, P. 2002. *Paradigm Lost: State Theory Reconsidered*. University of Minnesota Press.

Arrighi, G. and Silver, B. J. 1999a. Conclusion. Pages 271–289 of: *Chaos and Governance in the Modern World System*. University of Minnesota Press.

Arrighi, G. and Silver, B. J. 1999b. Introduction. Pages 1–36 of: *Chaos and Governance in the Modern World System*. University of Minnesota Press.

Arrighi, G. and Silver, B. J. 2001. Capitalism and world (dis)order. *Review of International Studies*, **27**, 257–279.

Ashby, W. R. 1960. *Design for a Brain: The Origin of Adaptive Behavior*. Chapman & Hall.

Babchuk, N., Bruce, K., and George, P. 1999. Collaboration in sociology and other scientific disciplines: a comparative trend analysis of scholarship in the social, physical, and mathematical sciences. *The American Sociologist*, **30**, 5–21.

Baker, W. E. 1992. The network organization in theory and practice. Pages 397–429 of: *Networks and Organizations*. Harvard Business School Press.

Balachandran, S., Kogut, B., and Harnal, H. 2010 (February). *The probability of default, excessive risk, and executive compensation: a study of financial services firms from 1995 to 2008*. Tech. rept. Columbia Business School Research Paper.

Balirwa, J. S., Chapman, C. A., Chapman, L. J., *et al.* 1953. Biodiversity and fishery sustainability in the Lake Victoria Basin: an unexpected marriage? *BioScience*, **2003**, 703–715.

Ball, F., Mollison, D., and Scalia-Tomba, G. 1997. Epidemics with two levels of mixing. *The Annals of Applied Probability*, **7**, 46–89.

Balthazard, P., Potter, R. E., and Warren, J. 2004. Expertise, extraversion and group interaction styles as performance indicators in virtual teams: how do perceptions of IT's performance get formed? *ACM SIGMIS Database*, **35**, 41–64.

Bandura, A. 1977. *Social Learning Theory*. General Learning Press.

Barabási, A. -L. 2003. *Linked: How Everything is Connected to Everything Else and What it Means for Business and Everyday Life*. Cambridge: Plume.

Barabási, A. -L. and Albert, R. 1999. Emergence of scaling in random networks. *Science*, **286**, 509–512.

Barabási, A. -L., Jeong, H., Néda, Z., *et al.* 2002. Evolution of the social network of scientific collaborations. *Physica A*, **311**, 590–614.

Barringer, B. R. and Harrison, J. S. 2000. Walking a tightrope: creating value through interorganizational relationships. *Journal of Management*, **26**, 367–403.

Barsade, S. G., Ward, A. J., Turner, J. D. F., and Sonnenfeld, J. A. 2000. To your heart's content: a model of affective diversity in top management teams. *Admistrative Science Quarterly*, **45**, 802–836.

Bascompte, J. 2007. Networks in ecology. *Basic and Applied Ecology*, **8**, 485–490.

Baum, J. A. C. and Oliver, C. 1991. Institutional linkages and organizational mortality. *Administrative Science Quarterly*, **36**, 187–218.

Baum, J. A. C., McEvily, B., and Rowley, T. 2007. Better with age: the longevity and the performance implications of bridging and closure. *Rotman School of Management, University of Toronto*, **Working paper n. 1032282**.

Bebchuk, L. A. and Grinstein, Y. 2005. The growth of executive pay. *Oxford Review of Economic Policy*, **21**, 283–303.

Bebchuk, L. A. and Spamann, H. 2010. Regulating bankers' pay. *Georgetown Law Journal*, **98**, 247–287.

Becker, G. S. 1968. Crime and punishment: an economic approach. *Journal of Political Economy*, **76**, 169–217.

Bennani-Chraïbi, M. and Fillieule, O. 2003. *Résistances et Protestations dans les Sociétés Musulmanes*. Presses de Sciences Po.

Bettencourt, L. M. A., Lobo, J., Helbing, D., Kühnert, C., and West, G. B. 1995. Explaining the level of bridewealth. *Current Anthropology*, **35**, 311–316.

Bettencourt, L. M. A., Lobo, J., Helbing, D., Kühnert, C., and West, G. B. 2007. Growth, innovation, scaling, and the pace of life in cities. *Proceedings of the National Academy of Sciences U.S.A.*, **104**, 7301–7306.

Bodini, A. 1990. What is the role of predation on stability of natural communities? A theoretical investigation. *Biosystems*, **26**, 21–30.

Bodini, A. 1998. Representing ecosystem structure through signed digraphs. Model reconstruction, qualitative predictions and management: the case of a freshwater ecosystem. *Oikos*, **83**, 93–106.

Bodini, A. 2000. Reconstructing trophic interactions as a tool for understanding and managing ecosystems: application to a shallow eutrophic lake. *Canadian Journal of Fisheries and Aquatic Sciences*, **57**, 1999–2009.

Bodini, A. and Bondavalli, C. 2002. Towards a sustainable use of water resources: a whole-ecosystem approach using network analysis. *International Journal of Environment and Pollution*, **18**, 463–485.

Bodini, A., Giavelli, G., and Rossi, O. 1994. The qualitative analysis of community food webs: implications for wildlife management and conservation. *Journal of Environmental Management*, **41**, 49–65.

Bodini, A., Bondavalli, C., and Allesina, S. 2007. *L'ecosistema e le sue relazioni. Idee e strumenti per la valutazione di impatto ambientale e di incidenza.* (in Italian). Franco Angeli Editore.

Bodini, A., Bellingeri, M., Allesina, S., and Bondavalli, C. 2009. Using food web dominator trees to catch secondary extinctions in action. *Philosophical Transactions of the Royal Society of London - Series B: Biological Sciences*, **364**, 1725–1731.

Bolton, P. and Dewatripont, M. 2005. *Contract Theory*. MIT Press.

Bondavalli, C., Bodini, A., Rossetti, G., and Allesina, S. 2006. Detecting stress at the whole ecosystem level. The case of a mountain lake: Lake Santo (Italy). *Ecosystems*, **9**, 768–787.

Borgatti, S. P. and Foster, P. 2003. The network paradigm in organizational research: a review and typology. *Journal of Management*, **29**, 991–1013.

Borgatti, S. P., Everett, M. G., and Freeman, L. C. 2002. *Ucinet for Windows: Software for Social Network Analysis*. Analytic Technologies, Harvard.

Borner, K., Maru, J. T., and Goldstone, R. L. 2004. The simultaneous evolution of author and paper networks. *Proceedings of the National Academy of Sciences U.S.A.*, **101**, 5266–5273.

Börner, K., Contractor, N. S., Falk-Krzesinski, H. J., *et al.* 2010. A multi-level systems perspective for the science of team science. *Science Translational Medicine*, **2**, 49cm24.

Boser, B. E., Guyon, I. M., and Vapnik, V. N. 1992. A training algorithm for optimal margin classifiers. Pages 144–152 of: *Proceedings of the 5th Annual Conference on Computational Learning Theory*. New York, NY, USA: ACM Press.

Bowers, R. V. 1937. The direction of intra-societal diffusion. *American Sociological Review*, **2**, 826–836.

Bowler, P. J. and Morus, I. R. 2005. *Making Modern Science: A Historical Survey*. Chicago: University of Chicago Press.

BP. 2008. *BP Statistical Review of World Energy 2008*. Tech. rept. British Petroleum.

Brass, D. J., Galaskiewicz, J., Greve, H. R., and Tsai, W. 2004. Taking stock of networks and organizations: a multilevel perspective. *Academy of Management Journal*, **47**, 795–817.

Braudel, F. 1973. *Capitalism and Material Life, 1400–1800*. Harper and Row.

Brauer, F. 2005. The Kermack McKendrick epidemic model revisited. *Mathematical Biosciences*, **198**, 119–131.

Breiger, R. L. 1974. The duality of persons and groups. *Social Forces*, **53**, 181–190.

Brown, J. R. 2000. Privatizing the university – the new tragedy of the commons. *Science*, **290**, 1701–1702.

Bunker, R. J. 2006. *Networks, Terrorism and Global Insurgency*. Routledge: London and New York.

Burstein, P. and Linton, A. 2002. The impact of political parties, interest groups, and social movement organizations on public policy. *Social Forces*, **75**, 135–169.

Burt, R. S. 1987. Social contagion and innovation: cohesion versus structural equivalence. *American Journal of Sociology*, **92**, 1287–1335.

Burt, R. S. 1992. *Structural Holes: The Social Structure of Competition*. Harvard University Press.

Burt, R. S. 1999. The social capital of opinion leaders. *Annals of the American Academy of Political and Social Science*, **566**, 37–54.

Burt, R. S. 2005. *Brokerage and Closure: An Introduction to Social Capital*. Oxford University Press.

Burt, R. S. 2008. Information and structural holes: comment on Reagans and Zuckerman. *Industrial and Corporate Change*, **17**, 953–969.

Callaway, D. S., Newman, M. E. J., Strogatz, S. H., and Watts, D. J. 2000. Network robustness and fragility: percolation on random graphs. *Physical Review Letters*, **85**, 5468–5471.

Camerano, L. 1880. Dell'equilibrio dei viventi mercé la reciproca distruzione. *Atti della Reale Accademia della Scienze di Torino*, **15**, 393–414.

Camerer, C. and Knez, M. 1996. Coordination, organizational boundaries and fads in business practices. *Industrial and Corporate Change*, **5**, 89–112.

Campbell, J. L. 2005. Where do we stand? Common mechanisms in organizations and social movement research. Pages 41–68 of: *Social Movements and Organizations*. Cambridge University Press.

Cardona, J. and Lacroix, C. 2008. *Chriffres Clés: Statistiques de la culture*. Paris: La Documentation française.

Carroll, W. K. and Ratner, R. S. 1996. Master framing and cross-movement networking in contemporary social movements. *Sociological Quarterly*, **37**, 601–625.

Carruthers, B. G. and Halliday, T. 1998. *Rescuing Business: The Making of Corporate Bankruptcy Law in England and the United States*. Oxford University Press.

Castells, M. 1996. *The Rise of the Network Society*. Blackwell.

Çetin, T. and Oguz, F. 2007. The reform in the Turkish natural gas market: a critical evaluation. *Energy Policy*, **35**, 3856–3867.

Ceballos, G. and Ehrlich, P. R. 2002. Mammal population losses and the extinction crisis. *Science*, **296**, 904–439.

Chandler, T. 1987. *Four Thousand Years of Urban Growth*. St. David's University Press.

Chang, C. -C. and Lin, C. -J. 2001. *LIBSVM: a library for support vector machines*. Software available at http://www.csie.ntu.edu.tw/~cjlin/libsvm.

Chaudhuri, K. N. 1978. *Trading World of Asia and the English East India Company*. Cambridge University Press.

Cheit, R. E. and Gersen, J. E. 2000. When businesses sue each other: an empirical study of state court litigation. *Law & Social Inquiry*, **25**, 789–816.

Chen, L. L., Blumm, N., Christakis, N. A., Barabási, A. -L., and Deisboeck, T. S. 2009. Cancer metastasis networks and the prediction of progression patterns. *British Journal of Cancer*, **101**, 749–758.

Christakis, N. A. and Fowler, J. H. 2009. *Connected: The Surprising Power of our Social Networks and how they Shape our Lives*. Little Brown and Co.

CIA. 2008. *The World Factbook*. Potomac Books.

Clark, H. H. and Wilkes-Gibbs, D. 1986. Referring as a collaborative process. *Cognition*, **22**, 1–39.

Clark, R. A. and Goldsmith, R. E. 2005. Market mavens: psychological influences. *Psychology and Marketing*, **22**, 289–312.

Cohen, J. E. 1994. Lorenzo Camerano's contribution to early food web theory. Pages 351–359 of: *Frontiers in Mathematical Biology, Part V: Frontiers in Community and Ecosystem Ecology*. Springer-Verlag.

Cohen, J. E., Briand, F., and Newman, C. M. 1990. *Community Food Webs: Data and Theory*. Springer–Verlag.

Cohen, M. D. and Bacdayan, E. 1994. Organizational routines are stored as procedural memory: evidence from a laboratory study. *Organization Science*, **5**, 554–568.

Cohen, R., Erez, K., Ben-Avraham, D., and Havlin, S. 2000a. Resilience of the Internet to Random Breakdowns. *Physical Review Letters*, **85**, 4626–4628.

Cohen, S., Brissette, I., Skoner, D., and Doyle, W. 2000b. Social integration and health: the case of the common cold. *Journal of Social Structure*, **1**, 1–7.

Cohen, W. and Levinthal, D. 1994. A fortune favors the prepared firm. *Management Science*, **40**, 227–251.

Coleman, J., Katz, E., and Menzel, H. 1957. The diffusion of an innovation among physicians. *Sociometry*, **20**, 253–270.

Collins, R. 1998. *The Sociology of Philosophies: A Global Theory of Intellectual Change*. Cambridge: Harvard University Press.

Compas, B. E., David, A., Haaga, F., *et al.* 1998. Sampling of empirically supported psychological treatments from health psychology: smoking, chronic pain, cancer, and bulimia nervosa. *Journal of Consulting and Clinical Psychology*, **66**, 89–112.

Contractor, N. S. 1994. New approaches to organizational communication. Pages 39–65 of: Kovacic, B. (ed), *New Approaches to Organizational Communication*. SUNY Press.

Contractor, N. S. 2009. The emergence of multidimensional networks. *Journal of Computer-Mediated Communication*, **14**, 743–747.

Contractor, N. S., Monge, P., and Leonardi, P. 2011. Multidimensional networks and the dynamics of sociomateriality: bringing technology inside the network. *International Journal of Communication*, **5**, 1–20.

Corden, W. M. and Neary, J. P. 1982. Booming sector and de-industrialisation in a small open economy. *Economic Journal*, **92**, 825–848.

Costanza, R., d'Arge, R., de Groot, R., *et al.* 1997. The value of the world's ecosystem services and natural capital. *Nature*, **387**, 253–260.

Coyle, D. 2011. *The Economics of Enough: How to Run the Economy as If the Future Matters*. Princeton University Press.

Cronin, B., Shaw, D. D., and La Barre, K. 2003. A cast of thousands: coauthorship and subauthorship collaboration in the 20th century as manifested in the scholarly journal literature of psychology and philosophy. *Journal of American Society for Information Technology*, **54**, 855–871.

Csárdi, G. and Nepusz, T. 2006. The igraph software package for complex network research. *InterJournal*, Complex Systems, 1695. http://igraph.sf.net.

Csermely, P. 2008. Creative elements: network-based predictions of active centres in proteins, cellular and social networks. *Trends in Biochemical Sciences*, **33**, 569–576.

Csermely, P. 2009. *Weak Links: The Universal Key to the Stability of Networks and Complex Systems*. Springer Verlag.

Cummings, J. N. and Kiesler, S. 2007. Who works with whom? Collaborative tie strength in distributed interdisciplinary projects. In: *Third International e-Social Science Conference*.

Dalton, R. 1994. *The Green Rainbow: Environmental Groups in Western Europe*. Yale University Press.

Dalton, R. 2008. *Citizen Politics*. 5th edn. CQ Press.

de Solla Price, D. J. 1963. *Little Science, Big Science ... and Beyond*. New York: Columbia University Press.

Degenne, A. and Forsé, M. 1999. *Introducing Social Networks*. Sage.

Dekker, A. H. 2007. Realistic social networks for simulation using network rewiring. Pages 677–683 of: *International Congress on Modelling and Simulation*.

Della Porta, D. 1995. *Social Movements, Political Violence and the State*. Cambridge University Press.

Dempster, A. P., Laird, N. M., and Rubin, D. B. 1977. Maximum likelihood from incomplete data via the EM algorithm. *Journal of the Royal Statistical Society, Series B*, **39**, 1–38.

Derényi, I., Farkas, I., Palla, G., and Vicsek, T. 2004. Topological phase transitions of random networks. *Physica A*, **334**, 583–590.

Devezas, T. 2001. The biological determinants of long wave behavior in socioeconomic growth and development. *Technological Forecasting and Social Change*, **68**, 1–57.

Devezas, T. and Modelski, G. T. 2008. The Portuguese as system-builders: technological innovation in early globalization. Pages 30–57 of: *Globalization as Evolutionary Process: Modeling, Simulating, and Forecasting Global Change*. Routledge.

Dewey, J. 1998. The pattern of inquiry. Pages 169–179 of: *The Essential Dewey, Volume 2: Ethics, Logic, Psychology*. Indiana University Press.

Di Battista, G., Eades, P., Tamassia, R., and Tollis, I. G. 1999. *Graph Drawing: Algorithms for the Visualization of Graphs*. Prentice Hall.

Diani, M. 1995. *Green Networks. A Structural Analysis of the Italian Environmental Movement*. Edinburgh University Press.

Diani, M. 2002. Network analysis. Pages 173–200 of: *Methods of Social Movement Research*. University of Minnesota Press.

Diani, M. and Bison, I. 2004. Organizations, coalitions, and movements. *Theory and Society*, **33**, 281–309.

DiMaggio, P. J. and Powell, W. W. 1983. The iron cage revisited: institutional isomorphism and collective rationality in organizational fields. *American Sociological Review*, **48**, 147–160.

DiPrete, T. A., Eirich, G., and Pittinsky, M. 2010. Compensation benchmarking, leapfrogs, and the surge in executive pay. *American Journal of Sociology*, **115**, 1671–1712.

Dunbar, R. I. M. 2005. Why are good writers so rare? An evolutionary perspective on literature. *Journal of Cultural and Evolutionary Psychology*, **3**, 7–22.

Dunne, J. A., Williams, R. J., and Martinez, N. D. 2002a. Food-web structure and network theory: The role of connectance and size. *Proceedings of the National Academy of Sciences U.S.A.*, **99**, 12917–12922.

Dunne, J. A, Williams, R. J., and Martinez, N. D. 2002b. Network structure and biodiversity loss in food webs: robustness increases with connectance. *Ecology Letters*, **5**, 558–567.

Dunne, J. A, Williams, R. J., and Martinez, N. D. 2004. Network structure and robustness of marine food webs. *Marine Ecology Progress Series*, **273**, 291–302.

Easley, D. and Kleinberg, J. 2010. *Networks, Crowds, and Markets: Reasoning About a Highly Connected World*. Cambridge University Press.

Ebenman, B. and Jonsson, T. 2005. Using community viability analysis to identify fragile systems and keystone species. *Trends in Ecology and Evolution*, **20**, 568–575.

Edmondson, A. 1999. Psychological safety and learning behavior in work teams. *Administrative Science Quarterly*, **44**, 350–383.

EEGA. 2008. *Major Gas Pipelines of the Former Soviet Union and Capacity of Export Pipelines*. Tech. rept. East European Gas Analysis. www.eegas.com/fsu.htm.

EIA. 2008. *Country Analysis Briefs*. Tech. rept. Energy Information Administration. www.eia.doe.gov/emeu/cabs.

English, J. F. 2005. *The Economy of Prestige: Prizes, Awards, and the Circulation of Cultural Value*. Cambridge: Harvard University Press.

Erdős, P. and Rényi, A. 1959. On random graphs. *Publicationes Mathematicae*, **6**, 290–297.

Erdős, P. and Rényi, A. 1960. On the evolution of random graphs. *Publications of the Mathematical Institute of the Hungarian Academy of Sciences*, **5**, 17–61.

Erikson, E. and Bearman, P. 2006. Foundations for global trade: the structure of English trade in the East Indies. *American Journal of Sociology*, **112**, 195–230.

Espinosa-Romero, M. J., Gregr, E. J., Walters, C., Christensen, V., and Chan, K. M. A. 2009. Representing mediating effects and species reintroductions in Ecopath with Ecosim. *Ecological Modelling*, **222**, 1569–1579.

Estes, J. A. and Palmisano, J. F. 1974. Sea otters: their role in structuring nearshore communities. *Science*, **185**, 1058–1060.

Etzkowitz, H., Kemelgor, C., Neuschatz, M., Uzzi, B., and Alonzo, J. 1994. The paradox of critical mass for women in science. *Science*, **266**, 51–54.

EU. 2002. *EU Green Paper: Towards a European Strategy for the Security of Energy Supply*. Tech. rept. European Commission.

Eurostat. 2008. *Energy – Yearly Statistics 2006, Edition 2008. Theme: Environment and energy Collection: Statistical books*. Tech. rept. European Union.

Evans, J. A. 2008. Electronic publication and the narrowing of science and scholarship. *Science*, **321**, 395–399.

Fafchamps, M., Goyal, S., and van der Leij, M. J. 2010. Matching and network effects. *Journal of the European Economic Association*, **8**, 203–231.

Fahlenbrach, R. and Stulz, R. M. 2011. Bank CEO incentives and the credit crisis. *Journal of Financial Economics*, **99**, 11–26.

Falconi, A. M., Guenfoud, K., Lazega, E., Lemercier, C., and Mounier, L. 2005. Le Contrôle social du monde des affaires: une étude institutionnelle. *L'Année Sociologique*, **55**, 451–484.

Falk-Krzesinski, H. J., Börner, K., Contractor, N. S., *et al.* 2010. Advancing the science of team science. *Clinical and Translational Sciences*, **3**, 263–266.

Falk-Krzesinski, H. J., Contractor, N. S., Fiore, S. M., *et al.* 2011. Mapping a research agenda for the science of team science. *Research Evaluation*, **20**, 143–156.

Faulkender, M. and Yang, J. 2010. Inside the black box: the role and composition of compensation peer groups. *Journal of Financial Economics*, **96**, 257–270.

Feick, L. F. and Price, L. L. 1987. The market maven: a diffuser of marketplace information. *The Journal of Marketing*, **51**, 83–97.

Fernandez-Mateo, I. 2007. Who pays the price of brokerage? Transferring constraint through price setting in the staffing sector. *American Sociological Review*, **72**, 291–317.

Festinger, L. and Schachter, S. 1950. *Social Pressures in Informal Groups: A Study of Human Factors in Housing*. Harper.

Figyelő. 2002. Kétszázak klubja 2001. In: *Figyelő Top 200*.

Fiore, S. M. 2008. Interdisciplinarity as teamwork how the science of teams can inform team science. *Small Group Research*, **39**, 251–277.

Fitzgerald, F. S. 1993. The Crack Up. Page 122 of: Wilson, E. (ed), *The Crack Up*. New York: New Directions.

Flanagin, A. J., Stohl, C., and Bimber, B. 2006. Modeling the structure of collective action. *Communication Monographs*, **73**, 29–54.

Flemming, R. B. 1998. Contested terrains and regime politics: thinking about America's trial courts and institutional change. *Law & Social Inquiry*, **23**, 941–965.

Fortunato, S. 2010. Community detection in graphs. *Physics Reports*, **486**, 75–174.

Freeman, L. C., Borgatti, S. P., and White, D. R. 1991. Centrality in valued graphs: A measure of betweenness based on network flow. *Social Networks*, **13**, 141–154.

Friedkin, N. E. 2004. Social cohesion. *Annual Review of Sociology*, **30**, 409–425.

Fruchterman, T. M. and Reingold, E. M. 1991. Graph drawing by force-directed placement. *Software – Practice and Experience*, **21**, 1129–1164.

Fulk, J., Flanagin, A., Kalman, M., Monge, P. R., and Ryan, T. 1996. Connective and communal public goods in interactive communication systems. *Communication Theory*, **61**, 60–87.

Fulk, J., Heino, R., Flanagin, A., Monge, P., and Bar, F. 2004. A test of the individual action model for organizational information commons. *Organization Science*, **15**, 569–585.

Gabaix, X. and Landier, A. 2008. Why has CEO pay increased so much? *The Quarterly Journal of Economics*, **123**, 49–100.

Galdwell, M. 2000. *The Tipping Point*. New York: Little Brown and Company.

Gamson, W. 1961. A theory of coalition formation. *American Sociological Review*, **26**, 373–382.

Gardiner, J. A. 2009. Defining corruption. Pages 25–40 of: Heidenheimer, A. J., and Johnston, M. (eds), *Political Corruption: Concepts and Contexts*, 5th edn. Transaction Publishers, New Brunswick, New Jersey.

Gasch, A. P., Spellman, P. T., Kao, C. M., *et al.* 2000. Genomic expression programs in the response of yeast cells to environmental changes. *Molecular Biology of the Cell*, **11**, 4241–4257.

George, E., Chattopadhyay, P., Sitkin, S. B., and Barden, J. 2006. Cognitive underpinnings of institutional persistence and change: a framing perspective. *Academy of Management Review*, **31**, 347–365.

Gergely, G., Derényi, I., Farkas, I., and Vicsek, T. 2005. Uncovering the overlapping community structure of complex networks in nature and society. *Nature*, **435**, 814–818.

Gerlach, L. 1971. Movements of revolutionary change. Some structural characteristics. *American Behavioral Scientist*, **43**, 813–836.

Gerlach, L. 2001. The structure of social movements: environmental activism and its opponents. Pages 289–310 of: *Networks and Netwars: The Future of Terror, Crime, and Militancy*. Rand.

Gerstner, L. V. 2002. *Who Says Elephants Can't Dance?* Harper Business, New York.

Giddens, A. 1994. *The Constitution of Society*. Polity Press.

GIE. 2008a. *Storage Map – Information by point*. Tech. rept. Gas Infrastructure Europe. http://www.gie.eu/maps_data/availablecapacities.asp.

GIE. 2008b. *The European Natural Gas Network (Capacities at cross-border points on the primary market)*. Tech. rept. Gas Infrastructure Europe. http://www.gie.eu/maps_data/capacity.asp.

GIE. 2010a. *LNG investment database*. Tech. rept. Gas Infrastructure Europe. www.gie.eu.com/maps_data/GLE/database.

GIE. 2010b. *LNG map – Information by point*. Tech. rept. Gas Infrastructure Europe. www.gie.eu.com/maps_data.

GIE. 2010c. *Storage investment database*. Tech. rept. Gas Infrastructure Europe. www.gie.eu/maps_data/GSE/database.

GIE. 2010d. *Storage map – Information by point*. Tech. rept. Gas Infrastructure Europe. www.gie.eu/maps_data.

Girvan, M. and Newman, M. E. J. 2002. Community structure in social and biological networks. *Proceedings of the National Academy of Sciences U.S.A.*, **99**, 7821–7826.

Glenn, J. K. 1999. Competing challengers and contested outcomes to state breakdown: the velvet revolution in Czechoslovakia. *Social Forces*, **78**, 187–211.

Goldenberg, J., Libai, B., and Muller, E. 2001. A complex systems look at the underlying process of word-of-mouth. *Marketing Letters*, **12**, 211–223.

Gordon, R. J. and Dew-Becker, I. 2008. *Controversies about the rise of American inequality: a survey.* Tech. rept. National Bureau of Economic Research Cambridge, Massachusetts, USA.

Gould, R. V. 1991. Multiple networks and mobilization in the Paris Commune. *American Sociological Review*, **56**, 716–729.

Gould, R. V. 2003. *Collision of Wills: How Ambiguity about Social Rank Breeds Conflict.* University of Chicago Press.

Granovetter, M. 1973. The strength of weak ties. *American Journal of Sociology*, **78**, 1360–1380.

Granovetter, M. 1978. Threshold models of collective behavior. *American Journal of Sociology*, **83**, 1420–1443.

Granovetter, M. 2005. Business groups and social organization. Pages 429–450 of: *Handbook of Economic Sociology.* Princeton University Press.

Granovetter, M. 2007. The social construction of corruption. In: Nee, V., and Swedberg, R. (eds), *On Capitalism.* Stanford University Press.

Green, S. and Green, K. 1996. *Broadway Musicals Show by Show.* 5th edn. Milwaukee: Hal Leonard.

Gruhl, D., Liben-Nowell, D., Guha, R., and Tomkins, A. 2004. Information diffusion through blogspace. *SIGKDD Explorations Newsletter*, **6**, 43–52.

Guimerà, R., Uzzi, B., Spiro, J., and Amaral, L. A. N. 2005. Team assembly mechanisms determine collaboration network structure and team performance. *Science*, **308**, 697–702.

Gulati, R. and Gargiulo, M. 1999. Where do interorganizational networks come from? *American Journal of Sociology*, **104**, 1439–1493.

Hackman, J. R. and Katz, N. 2010. Group behavior and performance. Pages 1208–1251 of: Fiske, S. T., Gilbert, D. T., and Lindzey, G. (eds), *Handbook of Social Psychology*, 5th edn. New York: Wiley.

Hahn, M. W. and Kern, A. D. 2005. Comparative genomics of centrality and essentiality in three eukaryotic protein-interaction networks. *Molecular Biology and Evolution*, **22**, 803–806.

Halbeisen, R. E. and Gerber, A. P. 2009. Stress-dependent coordination of transcriptome and translatome in yeast. *PLoS Biology*, **7**, e105.

Hall, B. H., Jaffe, A., and Trajtenberg, M. 2001. The NBER patent citations data file: lessons, insights and methodological tools. *National Bureau of Economic Research*, Working Paper No. 8498.

Hall, B. H., Jaffe, A. B., and Trajtenberg, M. 2005. Market value and patent citations. *RAND Journal of Economics*, **36**, 16–38.

Hannan, M. T. and Freeman, J. H. 1977. The population ecology of organizations. *American Journal of Sociology*, **82**, 929–964.

Hargreaves-Heap, S., Hollis, M., Lyons, B., Sugden, R., and Weale, A. 1992. *The Theory of Choice. A Critical Guide.* Blackwell.

Harrison, D. A., Price, K. H., and Bell, M. P. 1998. Beyond relational demography: time and the effects of surface- and deep-level diversity on work group cohesion. *Academy of Management Journal*, **41**, 96–107.

Hartwell, L. H., Hopfield, J. J., Leibler, S., and Murray, A. W. 1999. From molecular to modular cell biology. *Nature*, **402**, 47–52.

Hayek, F. A. 1945. The use of knowledge in society. *American Economic Review*, **35**, 519–530.

Heckman, J. J. 1976. The common structure of statistical models of truncation, sample selection and limited dependent variables and a simple estimator for such models. *Annals of Economic and Social Measurement*, **5**, 475–492.

Heckman, J. J. 1979. Sample selection bias as a specification error. *Econometrica*, **47**, 153–161.

Hellman, J. and Kaufmann, D. 2001. Confronting the challenge of state capture in transition economies. *Finance & Development*, **38**, 31–35.

Hellman, J., Jones, G., and Kaufmann, D. 2000. Seize the state, seize the day: state capture, corruption and influence in transition. *World Bank Policy Research*, **Working Paper No. 2444**.

Herman, I., Melançcon, G., and Scott Marshall, M. 2000. Graph visualization and navigation in information visualisation. *IEEE Transactions on Visualization and Computer Graphics*, **6**, 24–43.

Hethcote, H. W. 1989. Three basics epidemiological. *Biomathematics*, **18**, 119–144.

Hethcote, H. W. 2000. The Mathematics of Infectious Diseases. *SIAM Review*, **42**(4), 599–653.

Heydebrand, W. and Seron, C. 1990. *Rationalizing justice: The Political Economy of Federal District Courts*. SUNY Press.

Higashi, M., Burns, T. P., and Patten, B. C. 1989. Food network unfolding: an extension of trophic dynamics for application to natural ecosystems. *Journal of Theoretical Biology*, **140**, 243–261.

Hinds, P. J., Carley, K. M., Krackhardt, D., and Wholey, D. 2000. Choosing work group members: balancing similarity, competence, and familiarity. *Organizational Behavior and Human Decision Processes*, **81**, 226–251.

Holling, C. S. 1986. The resilience of terrestrial ecosystems: local surprise and global change. Pages 292–317 of: *Sustainable Development of the Biosphere*. Cambridge University Press.

Hollingshead, A. B. 1997. Retrieval processes in transactive memory systems. *Personality and Social Psychology*, **74**, 659–671.

Hollingshead, A. B. 1998. Communication, learning, and retrieval in transactive memory systems. *Journal of Experimental Social Psychology*, **34**, 423–442.

Holmgren, Å. J. 2006. Using graph models to analyze the vulnerability of electric power networks. *Risk Analysis*, **26**, 955–969.

Holstege, F. C., Jennings, E. G., Wyrick, J. J., *et al.* 1998. Dissecting the regulatory circuitry of a eukaryotic genome. *Cell*, **95**, 717–728.

Huang, M., Huang, Y., Ognyanova, K., *et al.* 2010. The effects of diversity and repeat collaboration on team performance in distributed nanoscientist teams. In: *Academy of Management Annual Conference*.

Huffaker, D. 2010. Dimensions of leadership and social influence in online communities. *Human Communication Research*, **36**, 593–617.

Ibolya, T. 2010. A hűtlen kezelés bizonyítása [Proving embezzlement]. *Rendészeti Szemle*, **10**, 82–94.

IEA. 1998. *Caspian Oil and Gas*. Tech. rept. International Energy Agency, Paris, France.

IEA. 2007a. *Energy balances of OECD countries 1960-2005*. Tech. rept. International Energy Agency, Paris, France.

IEA. 2007b. *World Energy Outlook 2007*. Tech. rept. OECD/IEA, Paris.

Jackson, B. A. 2006. Groups, networks, or movements: a command-and-control-driven approach to classifying terrorist organizations and its application to Al Qaeda. *Studies in Conflict and Terrorism*, **29**, 241–262.

Jacobs, J. 1969. *The Economy of Cities*. Random House.

Jacobs, J. 1983. The Economy of Regions. In: *Third Annual E. F. Schumacher Lecture, Mount Holyoke College, South Hadley*.

Jacobs, J. 1992. *Systems of Survival: A Dialogue on the Moral Foundations of Commerce and Politics*. Random House.

Jansen, J. C., van Arkel, W. G., and Boots, M. G. 2004. *Designing indicators of long-term energy supply security*. Tech. rept. ECN-C-04-007. Energy research Centre of the Netherlands. http://www.ecn.nl/docs/library/report/2004/c04007.pdf.

Jehn, K. A., Northcraft, G. B., and Neale, M. A. 1999. Why differences make a difference: a field study of diversity, conflict, and performance in workgroups. *Administrative Science Quarterly*, **44**, 741–763.

Jenkins, M. 2003. Prospects for biodiversity. *Science*, **302**, 1175–1177.

Jeong, H., Tombor, B., Albert, R., Oltvai, Z. N., and Barabási, A. -L. 2000. The large-scale organization of metabolic networks. *Nature*, **407**, 651–654.

Jeppesen, E., Kristensen, P., Jensen, J. P., Sçndergaard, M., Mortensen, E., and Lauridsen, T. 1991. Recovery resilience following a reduction in external phosphorous loading of shallow, eutrophic Danish lakes: duration, regulating factors and methods for overcoming resilience. *Memorie Istituto Italiano Idrobiologia*, **48**, 127–148.

Johnston, H. and Snow, D. A. 1998. Subcultures and the emergence of the Estonian nationalist opposition 1945-1990. *Sociological Perspectives*, **41**, 473–497.

Jones, B. F. 2005. The burden of knowledge and the death of the renaissance man: is innovation getting harder? *National Bureau of Economic Research*, **Working Paper No. 11360**.

Jones, C., Hesterly, W. S., and Borgatti, S. P. 1997. A general theory of network governance: exchange conditions and social mechanisms. *Academy of Management Review*, **22**, 911–945.

Jordán, F. 2009. Keystone species and food webs. *Philosophical Transactions of the Royal Society of London - Series B: Biological Sciences*, **364**, 1733–1741.

Jordán, F., Liu, W., and Wyatt, T. 2005. Topological constraints on the dynamics of wasp-waist ecosystems. *Journal of Marine Systems*, **57**, 250–263.

Kadushin, C., Ryan, D., Brodsky, A., and Saxe, L. 2005. Why it is so difficult to form effective community coalitions. *City and Community*, **4**, 255–275.

Kaplan, S. 2008. Are U.S. CEOs overpaid? *The Academy of Management Perspectives*, **22**, 5–20.

Kaplan, S. N. and Rauh, J. 2010. Wall Street and Main Street: what contributes to the rise in the highest incomes? *Review of Financial Studies*, **23**, 1004–1050.

Karl, T. L. 1997. *The Paradox of Plenty: Oil Booms and Petro-States*. University of California Press.

Katona, G. and Mueller, E. 1955. A study of purchase decisions. Pages 30–87 of: Clark, L. H. (ed), *Consumer Behavior: The Dynamics of Consumer Reaction*. New York University Press.

Katz, B. 2010. City centered. *Time*, **21**, 1–4.

Katz, E. 1996. Diffusion research at Columbia. Pages 61–70 of: Dennis, E. E., and Wartella, E. (eds), *American Communication Research: The Remembered History*. Lawrence Erlbaum Associates.

Katz, E. and Lazarsfeld, P. F. 1955. *Personal Influence: The Part Played by People in the Flow of Mass Communications*. New York: Free Press.

Katzenback, J. R. and Smith, D. K. 1993. *The Wisdom of Teams*. New York: Harper Business.

Kaufmann, D. and Kraay, A. 2007. On measuring governance: framing issues for debate. In: *Roundtable on Measuring Governance Hosted by the World Bank Institute and the Development Economics Vice-Presidency of The World Bank*.

Kaufmann, D., Kraay, A., and Mastruzzi, M. 2009. Governance matters VIII: aggregate and individual governance indicators, 1996-2008. *World Bank Policy Research*, **Working Paper No. 4978**.

Kempe, D., Kleinberg, J., and Tardos, É. 2003. Maximizing the spread of influence through a social network. Pages 137–146 of: *Proceedings of the 9th ACM SIGKDD International Conference on Knowledge Discovery and Data Mining*.

Kenis, P. and Knoke, D. 2002. How organizational field networks shape interorganizational tie-formation rates. *Academy of Management Review*, **27**, 275–293.

Khanna, T. and Rivkin, J. W. 2001. Estimating the Performance Effects of Business Groups in Emerging Markets. *Strategic Management Journal*, **22**, 45–74.

Kim, H. and Bearman, P. S. 1997. The structure and dynamics of movement participation. *Social Forces*, **62**, 70–93.

Kim, J., Kogut, B., and Yang, J.-S. 2011. *CEO pay, fat cats, and best athletes*. Tech. rept. Columbia Business School.

Klein, N. 2008. *The Shock Doctrine: The Rise of Disaster Capitalism*. Knopf.

Klitgaard, R. 1991. *Controlling Corruption*. University of California Press.

Knoke, D. 1990. *Political Networks*. Cambridge University Press.

Kogut, B. and Zander, U. 1992. Knowledge of the firm, combinative capabilities, and the replication of technology. *Organization Science*, **3**, 383–397.

Kolb, F. 2005. The impact of transnational protest on social movement organizations: mass media and the making of ATTAC Germany. Pages 95–120 of: Della Posta, D. and Tarrow, S. (eds), *Transnational Protest and Global Activism*. Rowman & Littlefield Publishers.

Kondoh, M. 2003. Foraging adaptation and the relationship between food-web complexity and stability. *Science*, **299**, 1388–1391.

Kontopoulos, K. 1993. *The Logics of Social Structure*. Cambridge University Press.

Korcsmáros, T., Kovács, I. A., Szalay, M. S., and Csermely, P. 2007. Molecular chaperones: the modular evolution of cellular networks. *Journal of Biosciences*, **32**, 441–446.

Kovács, I. A., Palotai, R., Szalay, M. S., and Csermely, P. 2010. Community landscapes: a novel, integrative approach for the determination of overlapping network modules. *PLoS ONE*, **7**, e12528.

Krackhardt, D. 1987. Cognitive social structures. *Social Networks*, **9**, 109–134.

Krackhardt, D. 1990. Assessing the political landscape: structure, cognition, and power in organizations. *Administrative Science Quarterly*, **35**, 342–369.

Krempel, L. and Plumper, T. 2003. Exploring the dynamics of international trade by combining the comparative advantages of multivariate statistics and network visualization. *Journal of Social Structure*, **4**, 1–22.

Kruyt, B., van Vuuren, D. P., de Vries, H. J. M., and Groenenberg, H. 2009. Indicators for energy security. *Energy Policy*, **37**, 2166–2181.

Kuhn, T. S. 1970. *The Structure of Scientific Revolutions*. Chicago: University of Chicago Press.

Kuran, T. 1987. Preference falsification, policy continuity and collective conservatism. *The Economic Journal*, **97**, 642–665.

Kynge, J. 2007. *China Shakes the World: a Titan's Rise and Troubled Future – and the Challenge for America*. New York: Houghton Mifflin.

Lambsdorff, J. G. 2007. *The Institutional Economics of Corruption and Reform Theory, Evidence and Policy*. Cambridge University Press.

Lammers, J. C. and Barbour, J. B. 2006. An institutional theory of organizational communication. *Communication Theory*, **16**, 356–377.

Larson, J. R., Christensen, C., Abbott, A. S., and Franz, T. M. 1996. Diagnosing groups: charting the flow of information in medical decision-making teams. *Journal of Personality and Social Psychology*, **71**, 315–330.

Lasswell, H. D. 1930. Bribery. In: Seligman, E. R. A. (ed), *Encyclopedia of Social Sciences*, vol. II. New York.

Laumann, E. and Knoke, D. 1987. *The Organizational State*. Wisconsin University Press.

Lawless, W. F., Rifkin, S., Sofge, D., *et al.* 2010. Conservation of information: reverse engineering dark social systems. *Structure and Dynamics: eJournal of Anthropological and Related Sciences*, **4**.

Lawrence, P. and Lorsch, J. 1967. Differentiation and integration in complex organizations. *Administrative Science Quarterly*, **12**, 1–30.

Lazarsfeld, P. F., Berelson, B., and Gaudet, H. 1948. *The People's Choice: How the Voter Makes Up His Mind in a Presidential Election*. New York: Columbia University Press.

Lazega, E. 1992. *Micropolitics of Knowledge: Communication and Indirect Control in Workgroups*. Aldine de Gruyter.

Lazega, E. 1994. Les conflits d'intérêts dans les cabinets américains d'avocats d'affaires: concurrence et auto-régulation. *Sociologie du Travail*, **36**, 315–336.

Lazega, E. 2003. *Networks in Legal Organizations: On the Protection of Public Interest in Joint Regulation of Markets*. Wiarda Institute Publications.

Lazega, E. 2009. Theory of cooperation among competitors: a neo–structural approach. *Sociologica*, **1**, 1–34.

Lazega, E. 2011. Four and half centuries of new (new) law and economics: legal pragmatism, shadow regulation, and institutional capture at the Commercial Court of Paris. In: de Vries, U., and Francot-Timmermans, L. (eds), *Law's Environment: Critical Legal Perspectives*. The Hague: Eleven International Publishing.

Lazega, E. and Mounier, L. 2002. Interdependent entrepreneurs and the social discipline of their cooperation: structural economic sociology for a society of organizations. In: *Conventions and Structures in Economic Organization: Markets, Networks, and Hierarchies*. Edward Elgar Publishers.

Lazega, E. and Mounier, L. 2003. Interlocking judges: on joint external and self–governance of markets. Pages 267–296 of: *Research in the Sociology of Organizations*. Emerald Group Publishing Limited.

Lazega, E., Mounier, L., Snijders, T., and Tubaro, P. in press. Norms, status and the dynamics of advice networks. *Social Networks,* **in press**.

Lazer, D. and Friedman, A. 2007. The network structure of exploration and exploitation. *Administrative Science Quarterly*, **52**, 667–694.

Lazer, D., Pentland, A., Adamic, L., *et al.* 2009. Social science: computational social science. *Science*, **323**, 721–723.

Lehner, J. 2010. A physicist solves the city. *NYT Magazine*.

Leitner, Y. 2005. Financial networks: contagion, commitment, and private sector bailouts. *The Journal of Finance*, **60**, 2925–2953.

Lengauer, T. and Tarjan, R. E. 1979. A fast algorithm for finding dominators in a flowgraph. *ACM Transactions on Programming Languages and Systems*, **1**, 121–141.

Lester, R. K. and Piore, M. J. 2004. *Innovation: The Missing Dimension*. Harvard University Press.

Levins, R. 1974. The qualitative analysis of partially specified systems. *Annals of New York Academy of Science*, **231**, 123–138.

Levins, R. 1995. Preparing for uncertainty. *Ecosystem Health*, **1**, 47–57.

Levy, S. F. and Siegal, M. L. 2008. Network hubs buffer environmental variation in Saccharomyces cerevisiae. *PLoS Biology*, **6**, e264.

Liljeros, F., Edling, C. R., Amaral, L. A. N., Stanley, H. E., and Åberg, Y. 2001. The web of human sexual contacts. *Nature*, **411**, 907–908.

Lincoln, J. R., Gerlach, M. L., and Ahmadjian, C. L. 1996. Keiretsu networks and corporate performance in Japan. *American Sociological Review*, **61**, 67–88.

Lochner, S. and Bothe, D. 2009. The development of natural gas supply costs to Europe, the United States and Japan in a globalizing gas market-Model-based analysis until 2030. *Energy Policy*, **37**, 1518–1528.

Lofland, J. 1996. *Social Movement Organizations*. Aldine de Gruyter.

Loreau, M., Naeem, S., Inchausti, P., *et al.* 2001. Biodiversity and ecosystem functioning: current knowledge and future challenges. *Science*, **294**, 804–808.

Luciani, G. 2004. *Security of Supply for Natural Gas Markets: What is it and what is it not?* Tech. rept. 119.04. FEEM. http://www.feem.it/userfiles/attach/Publication/NDL2004/NDL2004-119.pdf.

MacArthur, R. H. 1955. Fluctuation of animal populations and a measure of community stability. *Ecology*, **36**, 533–536.

Macaulay, S. 1963. Non-contractual relations in business: A preliminary study. *American Sociological Review*, **28**, 55–67.

Mahajan, V. and Peterson, R. A. 1985. *Models for Innovation Diffusion (Quantitative Applications in the Social Sciences*. Newbury Park, CA: Sage.

March, J. and Simon, H. 1958. *Organizations*. Wiley.

Marinova, E. 2010. *Natural gas demand and supply long term outlook to 2030 – ENTSOG TYNDP Workshop*.

Marsden, P. V. 2005. Recent developments in network measurement. Pages 8–30 of: *Models and Methods in Social Network Analysis*. Cambridge University Press.

Marshall, P. J. 1993. *The East India Company: A history*. Longman.

Martinez, N. D., Hawkins, B. A., Dawah, H. A., and Feifarek, B. P. 1999. Effect of sampling effort on characterization of food-web structure. *Ecology*, **80**, 1044–1055.

Marwell, G., Oliver, P. E., and Prahl, R. 1988. Social networks and collective action – a theory of the critical mass. *American Journal of Sociology*, **94**, 502–534.

Mavrakis, D., Thomaidis, F., and Ntroukas, I. 2006. An assessment of the natural gas supply potential of the south energy corridor from the Caspian Region to the EU. *Energy Policy*, **34**, 1671–1680.

May, R. M. 1973. *Complexity and Stability in Model Ecosystems*. Princeton University Press.

May, R. M. 2006. Network structure and the biology of populations. *Trends in Ecology and Evolution*, **21**, 394–399.

Mayntz, R. 2003. New challenges to governance theory. Pages 27–40 of: *Governance as Social and Political Communication*. Manchester University Press.

McAdam, D. 2003. Beyond structural analysis: toward a more dynamic understanding of social movements. Pages 281–298 of: *Social Movements and Networks*. Oxford University Press.

McCann, K. S. 2000. The diversity-stability debate. *Nature*, **405**, 228–233.

McDonald, K. 2002. From solidarity to fluidarity: social movements beyond "collective identity". The case of globalization conflicts. *Social Movement Studies*, **1**, 109–128.

McEvedy, C. and Jones, R. 1978. *Atlas of World Population History*. Penguin.

McIntosh, W. V. and Cates, C. L. 1997. *Judicial Entrepreneurship: The Role of the Judge in the Marketplace of Ideas*. Greenwood Press.

McPherson, M. and Smith-Lovin, L. 2002. Cohesion and membership duration: linking groups, relations and individuals in an ecology of affiliation. *Advances in Group Processes*, **19**, 1–36.

McPherson, M., Smith-Lovin, L., and Cook, J. M. 2001. Birds of a feather: homophily in social networks. *Annual Review of Sociology*, **27**, 415–444.

McVoy, E. C. 1940. Patterns of diffusion in the United States. *American Sociological Review*, **5**, 219–227.

Melián, C. and Bascompte, J. 2002. Complex networks: two ways to be robust? *Ecology Letters*, **5**, 705–708.

Melucci, A. 1996. *Challenging Codes*. Cambridge University Press.

Merton, R. K. 1968. The Matthew Effect in Science. *Science*, **159**, 56–63.

Merton, R. K. 1973a. *The Sociology of Science*. Chicago: University of Chicago Press. Pages 545–550.

Merton, R. K. 1973b. *The Sociology of Science*. Chicago: University of Chicago Press.

Metcalfe, B. and Linstead, A. 2003. Gendering teamwork: re-writing the feminine. *Gender, Work & Organization*, **10**, 94–119.

Meyer, D. S. and Corrigall-Brown, C. 2005. Coalitions and political context: U.S. movements against war in Iraq. *Mobilization*, **10**, 327–344.

Meyer, J. W. and Rowan, B. 1991. Institutionalized organizations: formal structure as myth and ceremony. Pages 41–62 of: Powell, W. W., and DiMaggio, P. J. (eds), *The New Institutionalism in Organizational Analysis*. Chicago: University of Chicago Press.

Mihalik, Á. and Csermely, P. 2011. Heat shock partially dissociates the overlapping modules of the yeast protein-protein interaction network: a systems level model of adaptation. *PLoS Computational Biology*, **7**, e1002187.

Milgram, S. 1967. The small-world problem. *Psychology Today*, **2**, 60–67.

Mizruchi, M. S. and Galaskiewicz, J. 1993. Networks of interorganizational relations. *Sociological Methods and Research*, **22**, 46–70.

Mizruchi, M. S. and Stearns, L. B. 1988. A Longitudinal study of the formation of interlocking directorates. *Administrative Science Quarterly*, **33**, 194–210.

Modelski, G. T. 1987. *Long Cycles in World Politics*. Seattle: University of Washington Press.

Modelski, G. T. 2000. What causes K-waves. *Technological Forecasting and Social Change*, **68**, 75–80.

Modelski, G. T. 2001. World system evolution. In: Denemark, R., Friedman, J., Gills, B., and T., Modelski G. (eds), *World-System Evolution: The Social Science of Long-Term Change*. Routledge.

Modelski, G. T. and Thompson, W. R. 1996. *Leading Sectors and World Powers: The Coevolution of Global Economics and Politics*. Columbia University Press.

Monge, P. R. and Contractor, N. S. 2003. *Theories of Communication Networks*. Oxford University Press.

Montoya, J. M. and Solé, R. V. 2002. Small world patterns in food webs. *Journal of Theoretical Biology*, **214**, 405–491.

Montoya, J. M., Pimm, S. L., and Solé, R. V. 2006. Ecological networks and their fragility. *Nature*, **214**, 259–493.

Moody, J. A. and Douglas, R. W. 2003. Structural cohesion and embeddedness: a hierarchical concept of social groups. *American Sociological Review*, **68**, 103–127.

Moreno, J. and Jennings, H. 1937. Statistics of social configurations. *Sociometry*, **1**, 342–374.

Moreno, J. L. 1934. *Who shall survive?: A new approach to the problem of human interrelations*. Nervous and Mental Disease Publishing Co., Washington, USA.

Navarrete, S. A. and Menge, B. A. 1996. Keystone predation and interaction strength: interactive effects of predators on their main prey. *Ecological Monographs*, **66**, 409–429.

Nelson, R. R. and Winter, S. G. 1983. *An Evolutionary Theory of Economic Change*. Boston: Bellknap.

Newman, M. E. J. 2001. The structure of scientific collaboration networks. *Proceedings of the National Academy of Sciences U.S.A.*, **98**, 404–409.

Newman, M. E. J. 2003. The Structure and function of complex networks. *SIAM Review*, **45**, 167–256.

Newman, M. E. J. 2004. Coauthorship networks and patterns of scientific collaboration. *Proceedings of the National Academy of Sciences U.S.A.*, **101**, 5200–5205.

Newman, M. E. J. 2006. Modularity and community structure in networks. *Proceedings of the National Academy of Sciences U.S.A.*, **103**, 8577–8582.

Newman, M. E. J. and Girvan, M. 2004. Finding and evaluating community structure in networks. *Physical Review E*, **69**, 026113.

Newman, M. E. J. and Watts, D. J. 1999. Scaling and percolation in the small-world network model. *Physical Review E*, **60**, 7332–7342.

Newman, M. E. J., Watts, D. J., and Strogatz, S. H. 2002. Random graph models of Social Networks. *Proceedings of the National Academy of Sciences U.S.A.*, **99**, 2566–2572.

Noack, A. 2007. Energy models for graph clustering. *Journal of Graph Algorithms and Applications*, **11**, 453–480.

Noack, A. and Rotta, R. 2009. Multi-level algorithms for modularity clustering. Pages 257–268 of: *SEA '09: Proceedings of the 8th International Symposium on Experimental Algorithms*. Springer-Verlag.

Nonaka, I. and Takeuchi, H. 1995. *The Knowledge-Creating Company*. Oxford University Press, New York.

Norris, P. 2003. *Democratic Phoenix*. Cambridge University Press.

Nye, J. S. 2008. Corruption and political development: a cost benefit analysis. Pages 281–300 of: Heidenheimer, A. J., and Johnston, M. (eds), *Political Corruption*. Transaction Publishers.

Nyírő, A. and Szakadát, I. 1993. *Politika Interaktív CD-rom*. Budapest: Aula.

Obstfeld, D. 2005. Social networks, the tertius iungens orientation, and involvement in innovation. *Administrative Science Quarterly*, **50**, 100–130.

Oka, R. C. and Chapurukha, M. K. 2008. Archaeology of trading systems, part 1: towards a new trade synthesis. *Journal of Archaeological Research*, **16**, 339–395.

Oka, R. C., Chapurukha, M. K., and Vishwas, D. G. 2009. Where others fear to trade: modeling adaptive resilience in ethnic trading networks to famines, maritime warfare and imperial stability in the growing Indian Ocean economy, ca. 1500–1700

CE. Pages 201–232 of: *The Political Economy of Hazards and Disasters*. Altamira Press.

Olcott, M. B. 2004. *International Gas Trade in Central Asia: Turkmenistan, Iran, Russia and Afghanistan*. Tech. rept. Working Paper #28. Stanford Institute for International Studies. www.bakerinstitute.org/publications.

Oliver, P. E. and Myers, D. J. 2003. Networks, diffusion, and cycles of collective action. Pages 173–204 of: Diani, M., and McAdam, D. (eds), *Social Movements and Networks*. Oxford: Oxford University Press.

Ouchi, W. G. 1980. Markets, bureaucracies and clans. *Administrative Science Quarterly*, **25**, 129–141.

Pagani, G. A. and Aiello, M. 2011. *The power grid as a complex network. Available online at: http://arxiv.org/abs/1105.3338v1*.

Paine, R. T. 1969. A note on trophic complexity and community stability. *American Naturalist*, **103**, 91–501.

Paine, R. T. 1980. Food webs, linkage interaction strength, and community infrastructure. *Journal of Animal Ecology*, **49**, 667–685.

Palla, G., Derényi, I., Farkas, I., and Vicsek, T. 2005. Uncovering the overlapping community structure of complex networks in nature and society. *Nature*, **435**, 814–818.

Palla, P., Barabási, A.-L., and Vicsek, T. 2007. Quantifying social group evolution. *Nature*, **466**, 664–667.

Palotai, R., Szalay, M. S., and Csermely, P. 2008. Chaperones as integrators of cellular networks: changes of cellular integrity in stress and diseases. *IUBMB Life*, **60**, 10–18.

Pascual, M. and Dunne, J. A. 2006. *Ecological Networks: Linking Structure to Dynamics in Food Webs*. Oxford University Press.

Pastor-Satorras, R. and Vespignani, A. 2001. Epidemic spreading in scale-free networks. *Physical Review Letters*, **86**, 3200–3203.

Pauly, D., Christensen, V., and Walters, C. 2000. Ecopath, Ecosim, and Ecospace as tools for evaluating ecosystem impact of fisheries. *ICES Journal of Marine Science*, **57**, 697–706.

Pavan, E. 2012. *Ties (In)Formation and Communication. Online and Offline Collaboration in the Internet Governance Forum Space*. Lanham: Rowman & Littlefield.

Pavlopoulos, G. A., Wegener, A. L., and Schneider, R. 2008. A survey of visualization tools for biological network analysis. *BioData Mining*, **1**, 12.

Pelletier, C. and Wortmann, J. C. 2009. A risk analysis for gas transport network planning expansion under regulatory uncertainty in Western Europe. *Energy Policy*, **37**, 721–732.

Peuhkuri, T. and Jokinen, P. 1999. The role of knowledge and spatial contexts in biodiversity policies: a sociological perspective. *Biodiversity and Conservation*, **8**, 133–147.

Pfeffer, J. and Salancik, J. 1978. *The External Control of Organizations*. New York: Harper and Row.

Piattoni, S. 2010. *The Theory of Multi-level Governance*. Oxford University Press.

Pimm, S. L., Lawton, J. H., and Cohen, J. E. 1991. Food web patterns and their consequences. *Nature*, **350**, 669–674.

Pimm, S. L., Raven, R., Peterson, A., Ekercioglu, C. H., and Ehrlich, P. R. 2006. Human impacts on the rates of recent, present, and future bird extinctions. *Proceedings of the National Academy of Sciences U.S.A.*, **103**, 10941–10946.

Pincus, S. C. A. 2005. *England's Glorious Revolution 1688–89: A Brief History with Documents*. St. Martin's Press.

Pirani, S., Stern, J., and Yafimava, K. 2009. *The Russo-Ukrainian gas dispute of January 2009: a comprehensive assessment*. Tech. rept. NG 27. Oxford Institute for Energy Studies.

Pizzorno, A. 1978. Political exchange and collective identity in industrial conflict. Pages 277–298 of: *The Resurgence of Class Conflict in Western Europe*. Holmes and Meier.

Pizzorno, A. 2008. *Rationality and Recognition*. Cambridge University Press.

Podolny, J. M. and Page, K. L. 1998. Network Forms of Organization. *Annual Review of Sociology*, **24**, 57–76.

Poole, M. S. and Contractor, N. S. 2011. Conceptualizing the multiteam system as a system of networked groups. Pages 193–224 of: Zaccaro, S. J., Marks, M. A., and DeChurch, L. A. (eds), *Multiteam Systems: An Organizational Form for Dynamic and Complex Environments*. Routledge Academic.

Poole, M. S. and DeSanctis, G. 1992. Microlevel structuration in computer-supported group decision-making. *Communication Research*, **19**, 5–49.

Poole, M. S., McPhee, R. D., and Seibold, D. R. 1982. A comparison of normative and interactional explanations of group decision-making: social decision schemes versus valence distributions. *Communication Monographs*, **49**, 1–19.

Poole, M. S., Seibold, D. R., and McPhee, R. D. 1996. The structuration of group decisions. Pages 114–146 of: Hirokawa, R., and Poole, M. (eds), *Communication and Group Decision Making*. Sage.

Postmes, T., Spears, R., and Lea, M. 2000. The formation of group norms in computer-mediated communication. *Human Communication Research*, **26**, 341–371.

Powell, W. 1990. Neither markets nor hierarchy: network forms of organization. *Research in Organizational Behavior*, **12**, 295–336.

Powell, W. W. and DiMaggio, P. J. 1991. *The New Institutionalism and Organizational Analysis*. University of Chicago Press.

Powell, W. W., White, D. R., Koput, K. W., and Jason, O. -S. 2005. Network dynamics and field evolution: the growth of interorganizational collaboration in the life sciences. *The American Journal of Sociology*, **110**, 1132–1205.

Proulx, S. R., Promislow, D. E., and Phillips, P. C. 2005. Network thinking in ecology and evolution. *Trends in Ecology and Evolution*, **20**, 345–353.

Provan, K. G. and Kenis, P. 2007. Modes of network governance: structure, management, and effectiveness. *Journal of Public Administration Research and Theory*, **18**, 229–252.

Provan, K. G., Fish, A., and Sydow, J. 2007. Interorganizational networks at the network level: A review of the empirical literature on whole networks. *Journal of Management*, **33**, 479–516.

Puccia, C. J. and Levins, R. 1985. *Qualitative Modelling of Complex Systems: An Introduction to Loop Analysis and Time Averaging*. Harvard University Press.

Puska, P. and Uutela, A. 2000. Community intervention in cardiovascular health promotion: North Karelia, 1972-1999. Pages 73–96 of: Schneiderman, N., Speers, M. A., Silva, J. M., Tomes, H., and Gentry, J. H. (eds), *Integrating Behavioral*

288 References

*and Social Sciences With Public Health.* Baltimore: United Book Press, American Psychological Association.

R Development Core Team. 2010. *R: A Language and Environment for Statistical Computing.* R Foundation for Statistical Computing, Vienna, Austria. http://www.R-project.org.

Ramasco, J. J., Dorogovtsev, S. N., and Pastor-Satorras, R. 2004. Self-organization of collaboration networks. *Physical Review E*, **70**, 036106.

Rasmusen, E. 1989. *Games and Information: An Introduction to Game Theory.* Blackwell Publising Ltd.

Ravasz, E., Somera, A. L., Mongru, D. A., Oltvai, Z. N., and Barabási, A. -L. 2002. Hierarchical organization of modularity in metabolic networks. *Science*, **297**, 1551–1555.

Reagans, R. and Zuckerman, E. W. 2001. Networks, diversity, and productivity: the social capital of corporate R&D teams. *Organization Science*, **12**, 502–517.

Reichardt, J. and White, D. R. 2007. Role models for complex networks. *European Physical Journal B*, **60**, 217–224.

Reymond, M. 2007. European key issues concerning natural gas: dependence and vulnerability. *Energy Policy*, **35**, 4169–4176.

Robins, G., Pattison, P., and Wang, P. 2006. Closure, connectivity and degrees: new specifications for Exponential Random Graph ($p*$) Models for directed social networks. *Unpublished manuscript. University of Melbourne.*

Robins, G., Pattison, P., and Wang, P. 2007. An introduction to Exponential Random Graph ($p*$) Models for social networks. *Social Networks*, **29**, 173–191.

Rogers, E. 2003. *Diffusion of Innovations.* Free Press, New York.

Rootes, C. 2003. *Environmental Protest in Western Europe.* Oxford University Press.

Rose-Ackerman, S. 1978. *Corruption A Study in Political Economy.* Academic Press.

Rose-Ackerman, S. 1999. *Corruption and Government Causes, Consequences, and Reform.* Cambridge University Press.

Rose-Ackerman, S. 2006. *International Handbook on the Economics of Corruption.* Edward Elgar.

Ruef, M., Aldrich, H. E., and Carter, N. M. 2003. The structure of founding teams: homophily, strong ties and isolation among U.S. entrepreneurs. *American Sociological Review*, **68**, 195–222.

Rutherford, S. L. and Lindquist, S. 1998. Hsp90 as a capacitor for morphological evolution. *Nature*, **396**, 336–342.

Rutledge, R. W., Basorre, B., and Mulholland, R. 1976. Ecological stability: an information theory viewpoint. *Journal of Theoretical Biology*, **57**, 355–371.

Ryan, B. and Gross, N. C. 1943. The diffusion of hybrid seed corn in two Iowa communities. *Rural Sociology*, **8**, 15–24.

Saavedra, S., Reed-Tsochas, F., and Uzzi, B. 2008. Asymmetric disassembly and robustness in declining networks. *Proceedings of the National Academy of Sciences U.S.A.*, **105**, 16466–16471.

Sampat, B. 2005. *Determinants of Patent Quality: An Empirical Analysis.* New York: Columbia University.

Sander, L. M., Warren, C. P., Sokolov, I. M., Simon, C., and Koopman, J. 2002. Percolation on heterogeneous networks as a model for epidemics. *Mathematical Biosciences*, **180**, 293–305.

Sartori, G. 1970. Concept misformation in comparative politics. *American Political Science Review*, **64**, 1033–1052.

Sassen, S. 1991. *The Global City: New York, London, Tokyo.* Princeton University Press.

Sassen, S. 2000. *Cities in a World Economy*. Pine Forge Press.

Sassen, S. 2009. Bridging the ecologies of cities and of nature. In: *The New Urban Question – Urbanism Beyond Neo-Liberalism. The 4th International Conference of the International Forum on Urbanism (IFoU)*.

Schaeffer, S. E. 2007. Graph Clustering. *Computer Science Review*, **1**, 27–64.

Schelling, T. C. 1978. *Micromotives and Macrobehavior*. New York: W. W. Norton.

Schilling, M. A. and Steensma, H. K. 2001. The use of modular organizational forms: an industry level analysis. *Academy of Management Journal*, **44**, 1149–1168.

Schumpeter, J. A. 1934. *The Theory of Economic Development*. Cambridge, Mass: Harvard University Press.

Schumpeter, J. A. 1942. *Capitalism, Socialism and Democracy*. Harper and Brothers.

Schwab, A. and Miner, A. S. 2008. Learning in hybrid-project systems: the effects of project performance on repeated collaboration. *Academy of Management Journal*, **51**, 1117–1149.

Schweitzer, F., Sornette, D., Vespignani, A., *et al.* 2009a. Economic networks: the new challenges. *Science*, **325**, 422–425.

Schweitzer, F., Sornette, D., Fagiolo, G., Vega-Redondo, F., and White, D. R. 2009b. Economic networks: what do we know and what do we need to know? *Advances in Complex Systems*, **12**, 407–422.

Scotti, M. 2008. Development Capacity. Pages 911–920 of: Jçrgensen, S. E., and Fath, B. D. (eds), *Ecological Indicators. Vol. [2] of Encyclopedia of Ecology*. Elsevier, Oxford.

Scotti, M., Bondavalli, C., and Bodini, A. 2009. Linking trophic positions and flow structure constraints in ecological networks: Energy transfer efficiency or topology effect? *Ecological Modelling*, **220**, 3070–3080.

Seibold, D. R. and Meyers, R. A. 2007. Group argument: a structuration perspective and research program. *Small Group Research*, **38**, 312–336.

Sewell, W. H. J. 1992. A theory of structure: duality, agency and transformation. *American Journal of Sociology*, **98**, 1–29.

Shalizi, C. 2011. *Do city economies scale with population? Manuscript. Carnegie Mellon University*.

Shawe-Taylor, J. and Cristianini, N. 2004. *Kernel methods for pattern analysis*. Cambridge University Press.

Shea, S. and Basch, C. E. 1990. A review of five major community-based cardiovascular disease prevention programs. Part I: rationale, design, and theoretical framework. *American Journal of Health Promotion*, **4**, 203–213.

Shirkani, N. 2008. *Egyptian gas flows to Israel*. Tech. rept. Upstream Online. www.upstreamonline.com/live/article150348.ece.

Sőti, C., Sreedhar, A. S., and Csermely, P. 2003. Apoptosis, necrosis and cellular senescence: chaperone occupancy as a potential switch. *Aging Cell*, **2**, 39–45.

Simas, R. 1988. *The Musicals No One Came to See: A Guidebook to Four Decades of Musical-Comedy Casualties on Broadway, Off Broadway and in Out-Of-Town Tryout, 1943-1983*. New York: Garland.

Simmel, G. 1898. The persistence of social groups. *American Journal of Sociology*, **3**, 662–698.

Simmel, G. 1964. *Conflict and the Web of Group Affiliations*. New York: Free Press.

Simon, H. A. 1959. Theories of decision-making in economics and behavioral science. *American Economic Review*, **3**, 253–283.

Smith, D. A. and White, D. R. 1992. Structure and dynamics of the global economy: network analysis of international trade 1965–1980. *Social Forces*, **70**, 857–894.

Smith, J. 2005a. Building bridges or building walls? Explaining regionalization among transnational social movement organizations. *Mobilization*, **10**, 251–270.

Smith, J. 2005b. Globalization and transnational social movement organizations. Pages 226–248 of: *Social Movements and Organizations*. Cambridge University Press.

Solomonoff, R. and Rapoport, A. 1951. Connectivity of random nets. *Bulletin of Mathematical Biology*, **13**, 107–117.

Somers, M. R. 1994. The narrative constitution of identity: a relational and network approach. *Theory and Society*, **23**, 605–649.

Song, H. and Kezunovic, M. 2007. A new analysis method for early detection and prevention of cascading events. *Electric Power Systems Research*, **77**, 1132–1142.

Sçrensen, E. and Torfing, J. 2007. *Theories of Democratic Network Governance*. Edward Elgar.

Sorenson, O., Rivkin, J. W., and Fleming, L. 2006. Complexity, networks and knowledge flow. *Research Policy*, **35**, 994–1017.

Spicer, A., McDermott, G. A., and Kogut, B. 2000. Entrepreneurship and privatization in Central Europe: the tenuous balance between destruction and creation. *Academy of Management Review*, **25**, 630–649.

Spufford, P. 2002. *Power and Profit: The Merchant in Medieval Europe*. Thames and Hudson.

Spufford, P. 2006. From Antwerp to London: the decline of financial centres in Europe. *De Economist*, **154**, 143–175.

Stark, C., Breitkreutz, B. J., Chatr-Aryamontri, A., *et al.* 2011. The BioGRID Interaction Database: 2011 update. *Nucleic Acids Research*, **39**, D698–D704.

Stark, D. 1996. Recombinant property in East European capitalism. *American Journal of Sociology*, **101**, 993–1027.

Stark, D. 2009. *The Sense of Dissonance: Accounts of Worth in Economic Life*. Princeton University Press.

Stark, D. and Bruszt, L. 1998. *Postsocialist Pathways: Transforming Politics and Property in East Central Europe*. Cambridge University Press.

Stark, D. and Vedres, B. 2006. Social times of network spaces: network sequences and foreign investment in Hungary. *American Journal of Sociology*, **111**, 1367–1412.

Stark, D. and Vedres, B. in press. Political holes in the economy: the business network of partisan firms in Hungary. *American Sociological Review*, in press.

Stasser, G., Stewart, D. D., and Wittenbaum, G. M. 1995. Expert roles and information exchange during discussion: the importance of knowing who knows what. *Journal of Experimental Social Psychology*, **31**, 244–265.

Stern, J. 2002. *Security of European natural gas supplies: the impact of import dependence and liberalization*. Tech. rept. 36 pp. Royal Institute of International Affairs, London. www.isn.ethz.ch/isn/Digital-Library/Publications/Detail/?id=99313 &lng=en.

Stern, J. 2006. Natural gas security problems in Europe: the Russian–Ukrainian crisis of 2006. *Asia-Pacific Review*, **13**, 32–59.

Stokols, D., Hall, K. L., Taylor, B. K., and Moser, R. P. 2008. The science of team science: overview of the field and introduction to the supplement. *American Journal of Preventive Medicine*, **2**, S77–S89.

Strang, D. and Meyer, J. W. 1993. Institutional conditions for diffusion. *Theory and society*, **22**, 487–511.

Strang, D. and Soule, S. 1998. Diffusion in organizations and social movements: from hybrid corn to poison pills. *Annual Review of Sociology*, **24**, 265–290.

Streeck, W. and Schmitter, P. C. 1985. Community, market, state – and associations? The prospective contribution of interest governance to social order. *European Sociological Review*, **1**, 119–138.

Swedberg, R. 1993. Non-contractual relations in business: a preliminary study. *Journal of Institutional and Theoretical Economics*, **149**, 204–209.

Sydow, J., Schreyögg, G., and Koch, J. 2009. Organizational path dependence: opening the black box. *Academy of Management Review*, **34**, 689–709.

Szalay, M. S., Kovács, I. A., Korcsmáros, T., Böde, C., and Csermely, P. 2007. Stress-induced rearrangements of cellular networks: consequences for protection and drug design. *FEBS Letters*, **581**, 3675–3680.

Szántó, Z. 1999. Principals, agents, and clients. Review of the modern concept of corruption. *Innovation: The European Journal of Social Sciences*, **12**, 629–632.

Szántó, Z. 2009. Kontraszelekció és erkölcsi kockázat a politikában [Adverse selection and moral hazard in politics]. *Közgazdasági Szemle*, **6**, 563–571.

Szántó, Z., Tóth, I. J., Varga, S., and Cserpes, T. 2011. A korrupció típusai és médiareprezentációja Magyarországon (2001–2009) [Types of corruption and their media representation in Hungary (2001–2009)]. *Belügyi Szemle*, **11**, 1–38.

Thevenot, L. 2001. Organized complexity: conventions of coordination and the composition of economic arrangements. *European Journal of Social Theory*, **4**, 405–425.

Thomas, P. 2006. The communication rights in the information society (CRIS) campaign. Applying social movement theories to an analysis of global media reform. *Gazette*, **68**, 291–312.

Thye, S. R., Yoon, J., and Lawler, E. J. 2002. The theory of relational cohesion: review of a research program. *Advances in Group Processes*, **19**, 139–166.

Tilly, C. 1984. *Big Structures, Large Processes, Huge Comparisons*. Russell Sage Foundation: New York.

Tilly, C. 1996. *The Contentious French: Four Centuries of Popular Struggle*. Cambridge: Harvard University Press.

Tilly, C. 2005. *Identities, Boundaries and Social Ties*. Paradigm.

Tilly, C. and Tarrow, S. 2007. *Contentious Politics*. Paradigm.

Tondera, D., Grandemange, S., Jourdain, A., *et al.* 2009. SLP-2 is required for stress-induced mitochondrial hyperfusion. *EMBO Journal*, **28**, 1589–1600.

Tóth, I. J. 2010. *A szélerőművek engedélyeztetésének labirintusa [Obtaining wind power plant permits]*. (Manuscript).

Trajtenberg, M. 1990. A penny for your quotes: patent citations and the value of innovations. *RAND Journal of Economics*, **21**, 172–187.

Turchin, P. 2003. *Historical Dynamics: Why States Rise and Fall*. Princeton University Press.

Turchin, P. 2005. Dynamical feedbacks between population growth and sociopolitical instability in agrarian states. *Structure and Dynamics*, **1**, 49–69.

Turchin, P. 2006. *War and Peace and War: The Rise and Fall of Empires*. Pi Press.

Turchin, P. 2009. A theory for formation of large empires. *Journal of Global History*, **4**, 191–217.

Turchin, P. 2010. Warfare and the evolution of social complexity: a multilevel-selection approach. *Structure and Dynamics*, **4**, 222–238.

Turchin, P. and Nefedov, S. 2009. *Secular Cycles*. Princeton University Press.

Tversky, A. and Kahneman, D. 1974. Judgment under uncertainty: heuristics and biases. *Science*, **185**, 1124–1131.

Ulanowicz, R. E. 1986. *Growth & Development: Ecosystems Phenomenology*. Springer Verlag, New York.

Ulanowicz, R. E. 1997. *Ecology, the Ascendent Perspective*. Columbia University Press.

Ulanowicz, R. E. 2004. Quantitative methods for ecological network analysis. *Computational Biology and Chemistry*, **28**, 321–339.

Ulanowicz, R. E. 2009. *A Third Window: Natural Life beyond Newton and Darwin*. Templeton Press.

Ulanowicz, R. E., Heymans, J. J., and Egnotovich, M. S. 2000. Network analysis of trophic dynamics in South Florida Ecosystems, FY 99: the Graminoid Ecosystem. *Annual Report to the United States Geological Service Biological Resources Division University of Miami Coral Gables, FL 33124*.

Useem, M. 1980. Corporations and the corporate elite. *Annual Review of Sociology*, **6**, 41–77.

Uzzi, B. 1997. Social structure and competition in interfirm networks: the paradox of embeddedness. *Administrative Science Quarterly*, **42**, 35–67.

Uzzi, B. and Spiro, J. 2005. Collaboration and creativity: the small world problem. *American Journal of Sociology*, **111**, 447–504.

Van Der Vegt, G. S., Bunderson, J. S., and Oosterhof, A. 2006. Expertness diversity and interpersonal helping in teams: why those who need the most help end up getting the least. *Academy of Management Journal*, **49**, 877–893.

Van Dyke, N. and McCammon, H. 2010. *Social Movement Coalitions*. University of Minnesota Press.

Vedres, B., and Stark, D. 2010. Structural folds: generative disruption in overlapping groups. *American Journal of Sociology*, **115**, 1150–1190.

Walgrave, S. and Rucht, D. 2010. *The World Says No to War: Demonstrations Against the War in Iraq*. University of Minnesota Press.

Wallerstein, I. 1974. *The Modern World-system. Vol. 1. Capitalist Agriculture and the Origin of the European World-economy in the Sixteenth Century*. Academic Press.

Wang, P., Sharpe, K., Robins, G. L., and Pattison, P. E. 2009. Exponential Random Graph ($p*$) Models for affiliation networks. *Social Networks*, **31**, 12–25.

Wang, S., Szalay, M. S., Zhang, C., and Csermely, P. 2008. Learning and innovative elements of strategy update rules expand cooperative network topologies. *PLoS ONE*, **3**, e1917.

Ward Jr., J. H. 1963. Hierarchical grouping to optimize an objective function. *Journal of the American Statistical Association*, **58**, 236–244.

Watts, D. J. 1999. Networks, dynamics and the small-world phenomenon. *American Journal of Sociology*, **105**, 493–527.

Watts, D. J. 2003. *Six Degrees: The Science of a Connected Age*. New York: W. W. Norton.

Watts, D. J. and Strogatz, S. H. 1998. Collective dynamics of "small-world" networks. *Nature*, **393**, 440–442.

WEC. 2007. *2007 Survey of Energy Resources*. Tech. rept. World Energy Council, London.

Wegner, D. M. 1987. Transactive memory: a contemporary analysis of the group mind. Pages 185–208 of: *Theories of Group Behavior*. Springer-Verlag.

Weick, K. E. 1992. *The Social Psychology of Organizing*. Addison–Wesley, Reading.

Weisser, H. 2007. The security of gas supply – a critical issue for Europe? *Energy Policy*, **35**, 1–5.

Weitzman, M. L. 1998. Recombinant growth. *Quarterly Journal of Economics*, **113**, 331–360.

Wellman, B. 1979. The community question: the intimate networks of East Yorkers. *American Journal of Sociology*, **84**, 1201–1231.

Wellman, B. and Gulia, M. 1999. Net surfers don't ride alone: virtual community as community. Pages 331–367 of: *Networks in the Global Village*. Westview Press.

Westman, W. E. 1990. Managing for biodiversity. *BioScience*, **30**, 26–33.

Wheeler, S., Mann, K., and Sarat, A. 1988. *Sitting in judgment: the sentencing of white collar criminals*. Yale University Press.

White, D. R. 2009a. *Dynamics of Human Behavior (Cohesion and Resistance)*. *Encyclopedia of Complexity and Systems Science*. Springer-Verlag.

White, D. R. 2009b. *The evolution of the medieval world economic network, and the Chinese link. Central European University, June Conference on "The Unexpected Link Using network science to tackle social problems," organized by Balázs Vedres & Marco Scotti, Inauguration of the CEU Center for Network Sciences. Powerpoint: http://intersci.ss.uci.edu/wiki/ppt/economic_networksCEU09.ppt – Old world city systems and economic networks 950-1950: how the growth and decline of cities and the rise and fall of city-size hierarchies is related to the network structure of intercity connections.*

White, D. R., Tambayong, L., and Kejžar, N. 2008. Oscillatory dynamics of city-size distributions in world historical systems. Pages 190–225 of: *Globalization as Evolutionary Process: Modeling, Simulating, and Forecasting Global Change*. Routledge.

White, D. R., Scott, D. W., Tolga, O., and F., Ren. 2011. *Exploratory causal analysis for networks of ethnographically well-studied populations*.

White, H. 1998. *Advances in Econometric Theory: The Selected Works of Halbert White*. Edward Elgar.

White, H. 2004. *New Perspectives in Econometric Theory: The Selected Works of Halbert White, Volume 2*. Edward Elgar.

White, H. C., Boorman, S. A., and Breiger, R. L. 1976. Social structure from multiple networks. I. Blockmodels of roles and positions. *American Journal of Sociology*, **81**, 730–780.

Williams, K. Y. and O'Reilly, C. A. 1998. Demography and diversity in organizations: a review of 40 years of research. Pages 77–140 of: *Research in Organizational Behavior*. JAI Press.

Wuchty, S., Jones, B., and Uzzi, B. 2007. The increasing dominance of teams in production of knowledge. *Science*, **316**, 1036–1039.

Yenikeyeff, S. M. 2008. *Kazakhstan's Gas: Export Markets and Export Routes*. Tech. rept. NG 25. Oxford Institute for Energy Studies. http://www.offnews.info/downloads/oxford_energy_kazakhstan_nov08.pdf.

Yinger, M. J. 1960. Contraculture and subculture. *American Sociological Review*, **25**, 625–635.

Zald, M. N. and McCarthy, J. D. 1980. Social movement industries: competition and cooperation among movement organizations. *Research in Social Movements, Conflict and Change*, **3**, 1–20.

Ziman, J. M. 1994. *Prometheus Bound*. Cambridge: Cambridge University Press.

Zucker, L. G. 1991. The role of institutionalism in cultural persistence. Pages 83–107 of: Powell, W. W., and DiMaggio, P. J. (eds), *The New Institutionalism in Organizational Analysis*. University of Chicago Press.

# Index